Industrial Waste Management

Industrial Waste Management

Edited by
Zander Ellis

Larsen & Keller
www.larsen-keller.com

Industrial Waste Management
Edited by Zander Ellis
ISBN: 978-1-63549-149-4 (Hardback)

▤ Larsen & Keller

Published by Larsen and Keller Education,
5 Penn Plaza,
19th Floor,
New York, NY 10001, USA

Cataloging-in-Publication Data

Industrial waste management / edited by Zander Ellis.
   p. cm.
Includes bibliographical references and index.
ISBN 978-1-63549-149-4
1. Factory and trade waste--Management. 2. Refuse and refuse Disposal. 3. Salvage (Waste, etc.).
4. Recycling (Waste, etc.). 5. Hazardous wastes--Management. I. Ellis, Zander.
TD897 .I63 2017
628.4--dc23

This book contains information obtained from authentic and highly regarded sources. All chapters are published with permission under the Creative Commons Attribution Share Alike License or equivalent. A wide variety of references are listed. Permissions and sources are indicated; for detailed attributions, please refer to the permissions page. Reasonable efforts have been made to publish reliable data and information, but the authors, editors and publisher cannot assume any responsibility for the vailidity of all materials or the consequences of their use.

Trademark Notice: All trademarks used herein are the property of their respective owners. The use of any trademark in this text does not vest in the author or publisher any trademark ownership rights in such trademarks, nor does the use of such trademarks imply any affiliation with or endorsement of this book by such owners.

The publisher's policy is to use permanent paper from mills that operate a sustainable forestry policy. Furthermore, the publisher ensures that the text paper and cover boards used have met acceptable environmental accreditation standards.

Printed and bound in the United States of America.

For more information regarding Larsen and Keller Education and its products, please visit the publisher's website www.larsen-keller.com

Table of Contents

Preface

Industrial waste is a major concern of environmentalists and government authorities around the world. Thus, the issue of managing this waste is a matter of concern. To combat this problem, countries and world organizations have passed some rules and regulations, which are followed by countries worldwide. Through this book, we attempt to provide the basic and fundamental concepts and laws of industrial waste management to students. The topics introduced in this extensive text are of utmost importance and provide deep knowledge about this area. As this subject is developing at a rapid pace, the contents of this textbook will help the students to understand the modern methods and applications of the subject.

Given below is the chapter wise description of the book:

Chapter 1- This chapter will provide an integrated understanding of industrial waste. Some of main causes of industrial waste are discussed in this chapter such as toxic waste, chemical waste, wastewater and suspended solids. This chapter is an overview of the subject matter incorporating all the major aspects of industrial waste.

Chapter 2- This chapter has been carefully written to provide an easy understanding of the varied aspects of industrial wastewater treatment. The processes and mechanisms used to treat wastewater expelled by industries comprise industrial wastewater treatments. Some of the techniques used for this are reverse osmosis, ion-exchange resin, flocculation and activated sludge.

Chapter 3- Tools and techniques are an important component of any field of study. Techniques such as incineration, pyrolysis and landfill are discussed in detail in this chapter. Incineration is a process that involves the burning of organic substances contained in waste materials whereas recycling is the method of converting waste into reusable objects.

Chapter 4- Radioactive waste contains radioactive material which is a standard outcome of nuclear power generation technologies. One of the distinctive features of radioactive waste is that it decays over a period of time; hence it needs to be isolated and managed in an appropriate place for an adequate period of time until it no longer poses a threat. This chapter provides a comprehensive overview of the subject.

Chapter 5- The converting of waste material into useful material in order to decrease the utilization of fresh raw materials is recycling. Battery recycling, paint recycling and concrete recycling are some of the recycling techniques contemplated in this chapter. Industrial waste treatment is best understood in confluence with the major topics listed in the following chapter.

Chapter 6- Oil spills are a common and extremely destructive form of man-made industrial accident. An API oil-water separator is a device designed to separate huge amounts of oil from the wastewater. There are other devices also used for separating oil such as the hydrocyclone. The chapter discusses the methods of oil separation in a critical manner providing key analysis to the subject matter.

Chapter 7- Conservation and resource reclamation is a developing field of study; the following chapter provides an outline, and also helps the reader to understand the subject of conservation. Some of the projects explained in this chapter are reclaimed water techniques, eco industrial park and industrial ecology.

Indeed, my job was extremely crucial and challenging as I had to ensure that every chapter is informative and structured in a student-friendly manner. I am thankful for the support provided by my family and colleagues during the completion of this book.

Editor

Introduction to Industrial Waste

This chapter will provide an integrated understanding of industrial waste. Some of main causes of industrial waste are discussed in this chapter such as toxic waste, chemical waste, wastewater and suspended solids. This chapter is an overview of the subject matter incorporating all the major aspects of industrial waste.

Industrial Waste

Industrial waste is the waste produced by industrial activity which includes any material that is rendered useless during a manufacturing process such as that of factories, industries, mills, and mining operations. It has existed since the start of the Industrial Revolution. Some examples of industrial wastes are chemical solvents, paints, sandpaper, paper products, industrial by-products, metals, and radioactive wastes.

Toxic waste, chemical waste, industrial solid waste and municipal solid waste are designations of industrial wastes. Sewage treatment plants can treat some industrial wastes, i.e. those consisting of conventional pollutants such as biochemical oxygen demand (BOD). Industrial wastes containing toxic pollutants require specialized treatment systems.

Management

In Thailand

In Thailand the roles in Municipal solid waste (MSW) management and industrial waste management are organized by the Royal Thai Government which is then divided into central government, regional government, and local government. Each government is responsible for different tasks. The central government is responsible to stimulate regulation, policies, and standards. The regional governments are responsible for coordinating the central and local governments. The local governments are responsible for waste management in their governed area. However, the local governments do not dispose of the waste by themselves but instead hire private companies that have been granted the right from the Pollution Control Department (PCD) in Thailand. The main companies are Bangpoo Industrial Waste Management Center, General Environmental Conservation Public Company Limited (GENCO), SGS Thailand, Waste Management Siam LTD (WMS), and Better World Green Public Company Limited (BWG). These companies are responsible for the waste they have received from their customers before releasing it to the environment, burying it, or using it for energy.

Toxic Waste

Toxic waste is any material in liquid, solid, or gas form that can cause harm by being inhaled, swallowed, or absorbed through the skin. Many of today's household products such as televisions, computers and phones contain toxic chemicals that can pollute the air and contaminate soils and water. Disposing of such waste is a major public health issue.

Valley of the Drums, a toxic waste site in Kentucky, United States, 1980.

Classifying Toxic Materials

Toxic materials are poisonous byproducts as a result of industries such as manufacturing, farming, construction, automotive, laboratories, and hospitals which may contain chemicals, heavy metals, radiation, dangerous pathogens, or other toxins. Toxic waste has become more abundant since the industrial revolution, causing serious global health issues. Disposing of such waste has become even more critical with the addition of numerous technological advances containing toxic chemical components. Products such as cellular telephones, computers, televisions, and solar panels contain toxic chemicals that can harm the environment if not disposed of properly to prevent the pollution of the air and contamination of soils and water. A material is considered toxic when it causes death or harm by being inhaled, swallowed, or absorbed through the skin.

The waste can contain chemicals, heavy metals, radiation, dangerous pathogens, or other toxins. Even households generate hazardous waste from items such as batteries, used computer equipment, and leftover paints or pesticides. Toxic material can be either human-made and others are naturally occurring in the environment. Not all hazardous substances are considered toxic.

The United Nations Environment Programme (UNEP) has identified 11 key substances that pose a risk to human health:

- Arsenic: used in making electrical circuits, as an ingredient in pesticides, and as a wood preservative. It is classified as a carcinogen.

- Asbestos: is a material that was once used for the insulation of buildings, and some businesses are still using this material to manufacture roofing materials and brakes. Inhalation of asbestos fibers can lead to lung cancer and asbestosis.

- Cadmium: is found in batteries and plastics. It can be inhaled through cigarette smoke, or digested when included as a pigment in food. Exposure leads to lung damage, irritation of the digestive track, and kidney disease.

- Chromium: is used as brick lining for high-temperature industrial furnaces, as a solid metal used for making steel, and in chrome plating, manufacturing dyes and pigments, wood preserving, and leather tanning. It is known to cause cancer, and prolonged exposure can cause chronic bronchitis and damage lung tissue.

- Clinical wastes: such as syringes and medication bottles can spread pathogens and harmful microorganisms, leading to a variety of illnesses.

- Cyanide: a poison found in some pesticides and rodenticides. In large doses it can lead to paralysis, convulsions, and respiratory distress.

- Lead: is found in batteries, paints, and ammunition. When ingested or inhaled can cause harm to the nervous and reproductive systems, and kidneys.

- Mercury: used for dental fillings and batteries. It is also used in the production of chlorine gas. Exposure can lead to birth defects and kidney and brain damage

- PCBs, or polychlorinated biphenyls, are used in many manufacturing processes, by the utility industry, and in paints and sealants. Damage can occur through exposure, affecting the nervous, reproductive, and immune systems, as well as the liver.

- POPs, persistent organic pollutants. They are found in chemicals and pesticides, and may lead to nervous and reproductive system defects. They can bio-accumulate in the food chain or persist in the environment and be moved great distances through the atmosphere.

- Strong acids and alkalis used in manufacturing and industrial production. They can destroy tissue and cause internal damage to the body.

The most overlooked toxic and hazardous wastes are the household products in everyday homes that are improperly disposed of such as old batteries, pesticides, paint, and car oil. Toxic waste can be reactive, ignitable, and corrosive.These types of waste are regulated under the Resource Conservation and Recovery Act (RCRA)

- Reactive wastes are those that can cause explosions when heated, mixed with water or compressed. They can release toxic gases into the air. They are unstable even in normal conditions. An example is Lithium-Sulfur Batteries.

- oCorrosive wastes are liquids capable of corroding metal containers. These are acids or bases that has a PH level of less than or equal to 2 or greater than or equal to 12.5. An example is Battery Acid.

With the increasing worldwide technology there are more substances that are being considered toxic and harmful to human health. Some of this technology includes cell phones and computers. They have been given the name e-waste or EEE, which stands for Electrical and Electronic Equipment. This term is also used for goods such as refrigerators, toys, and washing machines. These items can contain toxic components inside which can break down into our water systems when discarded. The reduction in the cost of these goods has allowed for these items to be distributed globally without thought or consideration to managing the goods once they become ineffective or broken.

In the United States, the Environmental Protection Agency (EPA) and the state departments oversee the rules that regulate hazardous waste. The EPA requires that toxic waste be handled with special precautions and be disposed of in designated facilities around the country. Also, many cities in the United States have collection days where household toxic waste is gathered. Some materials that may not be accepted at regular landfills are ammunition, commercially generated waste, explosives/shock sensitive items, hypodermic needles/syringes, medical waste, radioactive materials, and smoke detectors.

Health Defects

Toxic wastes often contain carcinogens, and exposure to these by some route, such as leakage or evaporation from the storage, causes cancer to appear at increased frequency in exposed individuals. For example, a cluster of the rare blood cancer polycythemia vera was found around a toxic waste dump site in northeast Pennsylvania in 2008.

The Human & Ecological Risk Assessment Journal conducted a study which focused on the health of individuals living near municipal landfills to see if it would be as harmful as living near hazardous landfills. They conducted a 7-year study that specifically tested for 18 types of cancers to see if the participants had higher rates than those that don't live around landfills. They conducted this study in western Massachusetts within a 1-mile radius of the North Hampton Regional Landfill.

People encounter these toxins buried in the ground, in stream runoff, in groundwater that supplies drinking water, or in floodwaters, as happened after Hurricane Katrina.

Some toxins, such as mercury, persist in the environment and accumulate. As a result of the bioaccumulation of mercury in both freshwater and marine ecosystems, predatory fish are a significant source of mercury in human and animal diets.

Handling and Disposal

One of the biggest problems with today's toxic material is how to dispose of it properly. Decades ago it was dumped into streams, rivers and oceans or buried underground with landfills.The United States created the regulatory agency called the Environmental Protection Agency (EPA) that was enacted in 1976 for the purpose of regulating toxic material. There are standards and laws developed to protect people and the environment from future harm due to improper handling of toxic waste. The agriculture industry uses over 800,000 tons of pesticides worldwide annually that contaminates the soils and eventually infiltrates into the groundwater causing drinking water to contain toxic chemicals. The oceans can be polluted from the stormdrain runoff of these chemicals as well. Toxic waste in the form of petrolium oil can either spill into the oceans from pipe leaks or large ships, but it can also enter the oceans from everyday citizens dumping car oil into the rainstorm sewer systems. Disposal is the placement of waste into or on the land. Disposal facilities are usually designed to permanently contain the waste and prevent the release of harmful pollutants to the environment. The most common hazardous waste disposal practice is placement in a land disposal unit such as a landfill, surface impoundment, waste pile, land treatment unit, or injection well. Land disposal is subject to requirements under EPA's Land Disposal Restrictions Program.

Organic wastes can be destroyed by incineration at high temperatures; however, if the waste contains heavy metals or radioactive isotopes, these must be separated and stored, as they cannot be destroyed. The method of storage will seek to immobilize the toxic components of the waste, possibly through storage in sealed containers, inclusion in a stable medium such as glass or a cement mixture, or burial under an impermeable clay cap. Waste transporters and waste facilities may charge fees; consequently, improper methods of disposal may be used to avoid paying these fees. Where the handling of toxic waste is regulated, the improper disposal of toxic waste may be punishable by fines or prison terms. Burial sites for toxic waste and other contaminated brownfield land may eventually be used as greenspace or redeveloped for commercial or industrial use.

History of US Toxic Waste Regulation

Resource Conservation and Recovery Act (RCRA) Enforcement,. The Act gives the United States Environmental Protection Agency the authority to control the generation, transportation, treatment, storage, and disposal of hazardous waste The Resource Conservation and Recovery Act was followed by the Toxic Substances Control Act, which took effect on January 1, 1977. The Act authorized the EPA to secure information on all new and existing chemical substances, as well as to control any substances that were determined to cause unreasonable risk to public health or the environment.

The Superfund Act is another act administered by the EPA. It contains rules about cleaning up toxic waste that was dumped illegally.

There has been a long ongoing battle between communities and environmentalists versus governments and corporations about how strictly and how fairly the regulations and laws are written and enforced. That battle began in North Carolina in the late summer of 1979, as EPA's TSCA regulations were being implemented. In North Carolina, PCB-contaminated oil was deliberately dripped along rural Piedmont highways, creating the largest PCB spills in American history and a public health crisis that would have repercussions for generations to come. The PCB-contaminated material was eventually collected and buried in a landfill in Warren County, but citizens' opposition, including large public demonstrations, exposed the dangers of toxic waste, the fallibility of landfills then in use, and EPA regulations allowing landfills to be built on marginal, but politically acceptable sites.

Warren County citizens argued that the toxic waste landfill regulations were based on the fundamental assumption that the EPA's conceptual dry-tomb landfill would contain the toxic waste. This assumption informed the siting of toxic waste landfills and waivers to regulations that were included in EPA's *Federal Register*. For example, in 1978, the base of a major toxic waste landfill could be no closer than five feet from ground water, but this regulation and others could be waived. The waiver to the regulation concerning the distance between the base of a toxic waste landfill and groundwater allowed the base to be only a foot above ground water if the owner/operator of the facility could demonstrate to the EPA regional administrator that a leachate collection system could be installed and that there would be no hydraulic connection between the base of the landfill and groundwater. Citizens argued that the waivers to the siting regulations were discriminatory mechanisms facilitating the shift from scientific to political considerations concerning the siting decision and that in the South this would mean a discriminatory proliferation of dangerous waste management facilities in poor black and other minority communities. They also argued that the scientific consensus was that permanent containment could not be assured. As resistance to the siting of the PCB landfill in Warren County continued and studies revealed that EPA dry-tomb landfills were failing, EPA stated in its *Federal Register* that all landfills would eventually leak and should only be used as a stopgap measure.

Years of research and empirical knowledge of the failures of the Warren County PCB landfill led citizens of Warren County to conclude that the EPA's dry-tomb landfill design and regulations governing the disposal of toxic and hazardous waste were not based on sound science and adequate technology. Warren County's citizens concluded also that North Carolina's *1981 Waste Management Act* was scientifically and constitutionally unacceptable because it authorized the siting of toxic, hazardous and nuclear waste facilities prior to public hearings, preempted local authority over the siting of the facilities, and authorized the use of force if needed.

In the aftermath of the Warren County protests, the 1984 Federal Hazardous and Solid Waste Amendments to the Resource Conservation and Recovery Act focused on waste minimization and phasing out land disposal of hazardous waste as well as corrective action for releases of hazardous materials. Other measures included in the 1984 amendments included increased enforcement authority for EPA, more stringent hazardous waste management standards, and a comprehensive underground storage tank program.

The disposal of toxic waste continues to be a source of conflict in the U.S. Due to the hazards associated with toxic waste handling and disposal, communities often resist the siting of toxic waste landfills and other waste management facilities; however, determining where and how to dispose of waste is a necessary part of economic and environmental policy-making.

The issue of handling toxic waste has become a global problem as international trade has arisen out of the increasing toxic byproducts produced with the transfer of them to less developed countries. In 1995, the United Nations Commission on Human Rights began to notice the illicit dumping of toxic waste and assigned a Special Rapporteur to examine the human rights aspect to this issue (Commission resolution 1995/81). In September 2011, the Human Rights Council decided to strengthen the mandate to include the entire life-cycle of hazardous products from manufacturing to final destination (aka cradle to grave), as opposed to only movement and dumping of hazardous waste. The title of the Special Rapporteur has been changed to the "Special Rapporteur on the implications for human rights of the environmentally sound management and disposal of hazardous substances and wastes."(Human Rights Council 18/11). The Human Rights Council has further extended the scope of its mandates as of September 2012 due to the result of the dangerous implications occurring to persons advocating environmentally sound practices regarding the generation,management, handling, distribution and final disposal of hazardous and toxic materials to include the issue of the protection of the environmental human rights defenders.

Mapping of Toxic Waste in The United States

TOXMAP is a Geographic Information System (GIS) from the Division of Specialized Information Services of the United States National Library of Medicine (NLM) that uses maps of the United States to help users visually explore data from the United States Environmental Protection Agency's (EPA) Superfund and Toxics Release Inventory programs. TOXMAP is a resource funded by the US Federal Government. TOXMAP's chemical and environmental health information is taken from NLM's Toxicology Data Network (TOXNET) and PubMed, and from other authoritative sources.

Chemical Waste

Chemical waste is a waste that is made from harmful chemicals (mostly produced by large factories). Chemical waste may fall under regulations such as COSHH in the United Kingdom, or the Clean Water Act and Resource Conservation and Recovery Act in the United States. In the U.S., the Environmental Protection Agency (EPA) and the Occupational Safety and Health Administration (OSHA), as well as state and local regulations also regulate chemical use and disposal. Chemical waste may or may not be classed as hazardous waste. A chemical hazardous waste is a solid, liquid, or gaseous material that displays either a "Hazardous Characteristic" or is specifically "listed" by name as a hazardous waste. There are four characteristics chemical wastes may have to be considered as hazardous. These are Ignitability, Corrosivity, Reactivity, and Toxicity. This type of hazardous waste must be categorized as to its identity, constituents, and hazards so that it may be safely handled and managed. Chemical waste is a broad term and encompasses many types of materials. Consult the Material Safety Data Sheet (MSDS), Product Data Sheet or Label for a list of constituents. These sources should state wheather this chemical waste is a waste that needs special disposal.

Chemical Waste Bin (Chemobox)

Guidance for Disposal of Laboratory Chemical Wastes

In the laboratory, chemical wastes are usually segregated on-site into appropriate waste carboys, and disposed by a specialist contractor in order to meet safety, health, and legislative requirements.

Innocuous aqueous waste (such as solutions of sodium chloride) may be poured down the sink. Some chemicals are washed down with excess water. This includes: concentrated and dilute acids and alkalis, harmless soluble inorganic salts (all drying agents), alcohols containing salts, hypochlorite solutions, fine (tlc grade) silica and alumina. Aqueous waste containing toxic compounds are collected separately

Waste elemental mercury, spent acids and bases may be collected separately for recycling.

Chemical waste category that should be followed for proper packaging, labelling, and disposal of chemical waste.

Waste organic solvents are separated into chlorinated and non-chlorinated solvent waste. Chlorinated solvent waste is usually incinerated at high temperature to minimize the formation of dioxins. Non-chlorinated solvent waste can be burned for energy recovery.

In contrast to this, chemical materials on the "Red List" should never be washed down a drain. This list includes: compounds with transitional metals, biocides, cyanides, mineral oils and hydrocarbons, poisonous organosilicon compounds, metal phosphides, phosphorus element, and fluorides and nitrites.

Moreover, the Environmental Protection Agency (EPA) prohibits disposing certain materials down any UVM drain. Including flammable liquids, liquids capable of causing damage to wastewater facilities (this can be determined by the pH), highly viscous materials capable of causing an obstruction in the wastewater system, radioactive materials, materials that have or create a strong odor, wastewater capable of significantly raising the temperature of the system, and pharmaceuticals or endocrine disruptors.

Broken glassware are usually collected in plastic-lined cardboard boxes for landfilling. Due to contamination, they are usually not suitable for recycling. Similarly, used hypodermic needles are collected as sharps and are incinerated as medical waste.

Chemical Compatibility Guideline

Many chemicals may react adversely when combined. It's recommended that incompatible chemicals are stored in separate areas of the lab.

Acids should be separated from alkalis, metals, cyanides, sulfides, azides, phosphides, and oxidizers. The reason being, when combined acids with these type of compounds, violent exothermic reaction can occur possibly causing flammable gas, and in some cases explosions.

Oxidizers should be separated from acids, organic materials, metals, reducing agents, and ammonia. This is because when combined oxidizers with these type of compounds, inflammable, and sometimes toxic compounds can occur.

Container Compatibility

When disposing hazardous laboratory chemical waste, chemical compatibility must be considered. For safe disposal, the container must be chemically compatible with the material it will hold. Chemicals must not react with, weaken, or dissolve the container or lid. Acids or bases should not be stored in metal. Hydrofluoric acid should not store in glass. Gasoline (solvents) should not store or transport in lightweight polyethylene containers such as milk jugs. Moreover, the Chemical Compatibility Guidelines should be considered for more detailed information.

Laboratory Waste Containers

Packaging, labelling, storage are the three requirements for disposing chemical waste.

Packaging

filled up to 75% capacity to allow for vapour expansion and to reduce potential spills which could occur from moving overfilled containers. Container material must be compatible with the stored hazardous waste. Finally, wastes must not be packaged in containers that improperly identify other nonexisting hazards.

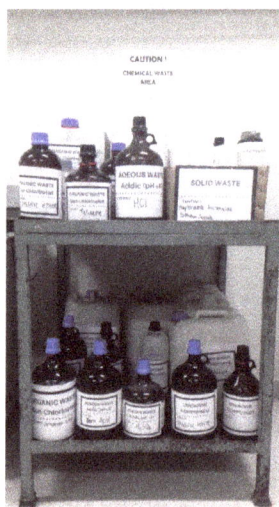

How to properly label, package, and store chemical waste safely.

For packaging, chemical liquid waste containers should only be In addition to the general packaging requirements mentioned above, incompatible materials should never be mixed together in a single container. Wastes must be stored in containers compatible with the chemicals stored as mentioned in the container compatibility section. Solvent safety cans should to be used to collect and temporarily store large volumes (10-20 litres) of flammable organic waste solvents, precipitates, solids or other non-fluid wastes should not be mixed into safety cans.

Labelling

Label all containers with the group name from the chemical waste category and an itemized list of the contents. All chemicals or anything contaminated with chemicals posing a significant hazard. All waste must be appropriately packaged.

Storage

When storing chemical wastes, the containers must be in good condition and should remain closed unless waste is being added. Hazardous waste must be stored safely prior to removal from the laboratory and should not be allowed to accumulate. Container should be sturdy and leakproof, also has to be labeled. All liquid waste must be stored in leakproof containers with a screw- top or other secure lid. Snap caps, mis-sized caps, parafilm and other loose fitting lids are not acceptable. If necessary, transfer waste material to a container that can be securely closed. Keep waste containers closed except when adding waste. Secondary containment should be in place to capture spills and leaks from the primary container, segregate incompatible hazardous wastes, such as acids and bases.

Mapping of Chemical Waste in The United States

TOXMAP is a Geographic Information System (GIS) from the Division of Specialized Information Services of the United States National Library of Medicine (NLM) that uses maps of the United States to help users visually explore data from the United States Environmental Protection Agency's (EPA) Toxics Release Inventory and Superfund Basic Research Programs. TOXMAP is a resource funded by the US Federal Government. TOXMAP's chemical and environmental health information is taken from NLM's Toxicology Data Network (TOXNET) and PubMed, and from other authoritative sources.

Chemical Waste in Canadian Aquaculture

Chemical waste in our oceans is becoming a major issue for the marine life. There have been many studies conducted to try and prove the effects of these chemical in our oceans. In Canada, many of the studies concentrated on the Atlantic provinces, where fishing and aquaculture are an important part of the economy. In New Brunswick, a study was done on the sea urchin in an attempt to identify the effects of toxic and chemical waste on life beneath the ocean, specifically the wasted from the salmon farms. Sea urchins were used to check the levels of metals in the environment. It is advantageous to use green sea urchins, Strongylocentrotus droebachiensis, because they are widely distributed, abundant in many locations, and easily accessible. By investigating the concentrations of metals in the green sea urchins, the impacts of produced chemicals from salmon aquaculture activity could be assessed and detected. Samples were taken at 25m intervals along a transect in the direction of the main tidal flow. The study found that there was impacts to at least 75m based on the intestine metal concentrations. So based on this study it is clear that the metals are contaminating the oceans and negatively affecting aquatic life.

Green Sea Urchin or S. droebacheinsis

Uranium in Ground and Surface Water in Canada

Another issue regarding chemical waste is the potential risk of surface and groundwater contamination by the heavy metals and radionuclides leached from uranium waste-rock piles (UWRP) A Radionuclide is an atom that has excess nuclear energy, making it unstable. Uranium waste-rock piles refers to Uranium mining, which is the process of extraction of uranium ore from the ground. . An example of such threats is in Saskatchewan, Uranium mining and ore processing (milling) can pose a threat to the environment. In open pit mining, large amounts of materials are excavated and disposed off in waste-rock piles. Waste-rock piles from the Uranium mining industry can contain several heavy metals and contaminants that may become mobile under certain conditions. Environmental contaminants may include acid mine drainage, higher concentrations of radionuclides, and non-radioactive metals/metalloids (i.e. As, Mo, Ni, Cu, Zn).

The leachability of heavy metals and radionuclide from UWRP plays a significant role in determining their potential environmental risks to surrounding surface and groundwater. Substantial differences in the solid-phase partitioning and chemical leachability of Ni and U were observed in the investigated UWRP lithological materials and background organic-rich lake sediment. For Instance, in the uranium-mining district of Northern Saskatchewan, Canada, the sequential extraction results showed that a significant amount of Ni (Nickel) was present in the non-labile residual fraction, while Uranium was mostly distributed in the moderately labile fractions. Although Nickel was much less labile than Uranium, the observed Nickel exceeded Uranium concentrations in leaching]].The observed Nickel and Uranium concentrations were relatively high in the underlying organic-rich lake sediment. Expressed as the percentage of total metal content, potential leachability decreased in the order U > Ni. Data suggest that these elements could potentially migrate to the water table below the UWRP. Detailed information regarding the solid-phase distribution of contaminants in the UWRP is critical to understand the potential for their environmental transport and mobility

The most visible civilian use of uranium is as the thermal power source used in nuclear power plants

Wastewater

Wastewater, also written as waste water, is any water that has been adversely affected in quality by anthropogenic influence. Wastewater can originate from a combination of domestic, industrial, commercial or agricultural activities, surface runoff or stormwater, and from sewer inflow or infiltration.

Wastewater treatment plant in Cuxhaven, Germany

Municipal wastewater (also called sewage) is usually conveyed in a combined sewer or sanitary sewer, and treated at a wastewater treatment plant. Treated wastewater is discharged into receiving water via an effluent pipe. Wastewaters generated in areas without access to centralized sewer systems rely on on-site wastewater systems. These typically comprise a septic tank, drain field, and optionally an on-site treatment unit. The management of wastewater belongs to the overarching term sanitation, just like the management of human excreta, solid waste and stormwater (drainage).

Sewage is a type of wastewater that comprises domestic wastewater and is therefore contaminated with feces or urine from people's toilets, but the term sewage is also used to mean any type of wastewater. Sewerage is the physical infrastructure, including pipes, pumps, screens, channels etc. used to convey sewage from its origin to the point of eventual treatment or disposal.

Origin

Wastewater can come from:

- Human excreta (feces and urine) often mixed with used toilet paper or wipes; this is known as blackwater if it is collected with flush toilets

- Washing water (personal, clothes, floors, dishes, cars, etc.), also known as greywater or sullage

- Surplus manufactured liquids from domestic sources (drinks, cooking oil, pesticides, lubricating oil, paint, cleaning liquids, etc.)

- Urban rainfall runoff from roads, carparks, roofs, sidewalks/pavements (contains oils, animal feces, litter, gasoline/petrol, diesel or rubber residues from tires, soapscum, metals from vehicle exhausts, etc.)

- Highway drainage (oil, de-icing agents, rubber residues, particularly from tires)

- Storm drains (may include trash)

- Manmade liquids (illegal disposal of pesticides, used oils, etc.)

- Industrial waste

- Industrial site drainage (silt, sand, alkali, oil, chemical residues);

    o Industrial cooling waters (biocides, heat, slimes, silt)

    o Industrial process waters

    o Organic or biodegradable waste, including waste from abattoirs, creameries, and ice cream manufacture

    o Organic or non bio-degradable/difficult-to-treat waste (pharmaceutical or pesticide manufacturing)

    o Extreme pH waste (from acid/alkali manufacturing, metal plating)

    o Toxic waste (metal plating, cyanide production, pesticide manufacturing, etc.)

    o Solids and emulsions (paper manufacturing, foodstuffs, lubricating and

> hydraulic oil manufacturing, etc.)

- o Agricultural drainage, direct and diffuse

- o Hydraulic fracturing

- o Produced water from oil & natural gas production

Wastewater can be diluted or mixed with other types of water in the form of:

- Seawater ingress (high volumes of salt and microbes)

- Direct ingress of river water

- Rainfall collected on roofs, yards, hard-standings, etc. (generally clean with traces of oils and fuel)

- Groundwater infiltrated into sewage

After it has undergone some treatment, the "treated wastewater" remains, e.g.:

- Septic tank discharge

- Sewage treatment plant discharge

Constituents

The composition of wastewater varies widely. This is a partial list of what it may contain:

- Water (more than 95 percent), which is often added during flushing to carry waste down a drain;

- Pathogens such as bacteria, viruses, prions and parasitic worms;

- Non-pathogenic bacteria;

- Organic particles such as feces, hairs, food, vomit, paper fibers, plant material, humus, etc.;

- Soluble organic material such as urea, fruit sugars, soluble proteins, drugs, pharmaceuticals, etc.;

- Inorganic particles such as sand, grit, metal particles, ceramics, etc.;

- Soluble inorganic material such as ammonia, road-salt, sea-salt, cyanide, hydrogen sulfide, thiocyanates, thiosulfates, etc.;

- Animals such as protozoa, insects, arthropods, small fish, etc.;

- Macro-solids such as sanitary napkins, nappies/diapers, condoms, needles, children's toys, dead animals or plants, etc.;

- Gases such as hydrogen sulfide, carbon dioxide, methane, etc.;

- Emulsions such as paints, adhesives, mayonnaise, hair colorants, emulsified oils, etc.;

- Toxins such as pesticides, poisons, herbicides, etc.

- Pharmaceuticals and hormones and other hazardous substances

Quality Indicators

Any oxidizable material present in an aerobic natural waterway or in an industrial wastewater will be oxidized both by biochemical (bacterial) or chemical processes. The result is that the oxygen content of the water will be decreased. Basically, the reaction for biochemical oxidation may be written as:

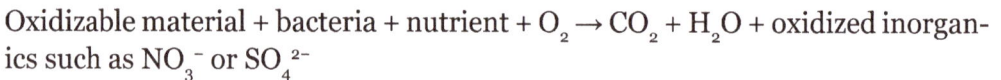

Oxidizable material + bacteria + nutrient + $O_2 \rightarrow CO_2$ + H_2O + oxidized inorganics such as NO_3^- or SO_4^{2-}

Oxygen consumption by reducing chemicals such as sulfides and nitrites is typified as follows:

$$S^{2-} + 2\,O_2 \rightarrow SO_4^{2-}$$

$$NO_2^- + \tfrac{1}{2}\,O_2 \rightarrow NO_3^-$$

Since all natural waterways contain bacteria and nutrients, almost any waste compounds introduced into such waterways will initiate biochemical reactions (such as shown above). Those biochemical reactions create what is measured in the laboratory as the biochemical oxygen demand (BOD). Such chemicals are also liable to be broken down using strong oxidizing agents and these chemical reactions create what is measured in the laboratory as the chemical oxygen demand (COD). Both the BOD and COD tests are a measure of the relative oxygen-depletion effect of a waste contaminant. Both have been widely adopted as a measure of pollution effect. The BOD test measures the oxygen demand of biodegradable pollutants whereas the COD test measures the oxygen demand of oxidizable pollutants.

The so-called 5-day BOD measures the amount of oxygen consumed by biochemical oxidation of waste contaminants in a 5-day period. The total amount of oxygen consumed when the biochemical reaction is allowed to proceed to completion is called the Ultimate BOD. Because the Ultimate BOD is so time consuming, the 5-day BOD has been almost universally adopted as a measure of relative pollution effect.

There are also many different COD tests of which the 4-hour COD is probably the most common.

There is no generalized correlation between the 5-day BOD and the ultimate BOD. Similarly there is no generalized correlation between BOD and COD. It is possible to develop such correlations for specific waste contaminants in a specific wastewater stream but such correlations cannot be generalized for use with any other waste contaminants or wastewater streams. This is because the composition of any wastewater stream is different. As an example an effluent consisting of a solution of simple sugars that might discharge from a confectionery factory is likely to have organic components that degrade very quickly. In such a case, the 5 day BOD and the ultimate BOD would be very similar since there would be very little organic material left after 5 days. However a final effluent of a sewage treatment works serving a large industrialised area might have a discharge where the ultimate BOD was much greater than the 5 day BOD because much of the easily degraded material would have been removed in the sewage treatment process and many industrial processes discharge difficult to degrade organic molecules.

The laboratory test procedures for the determining the above oxygen demands are detailed in many standard texts. American versions include the "Standard Methods for the Examination of Water and Wastewater."

Disposal

In some urban areas, municipal wastewater is carried separately in sanitary sewers and runoff from streets is carried in storm drains. Access to either of these is typically through a manhole. During high precipitation periods a combined sewer overflow can occur, forcing untreated sewage to flow back into the environment. This can pose a serious threat to public health and the surrounding environment.

Industrial wastewater effluent with neutralized pH from tailing runoff. Taken in Peru.

Sewage may drain directly into major watersheds with minimal or no treatment but this usually has serious impacts on the quality of an environment and on the health of people. Pathogens can cause a variety of illnesses. Some chemicals pose risks even at very low concentrations and can remain a threat for long periods of time because of bioaccumulation in animal or human tissue.

Treatment

There are numerous processes that can be used to clean up wastewaters depending on the type and extent of contamination. Wastewater can be treated in wastewater treatment plants which include physical, chemical and biological treatment processes. Municipal wastewater is treated in sewage treatment plants (which may also be referred to as wastewater treatment plants). Agricultural wastewater may be treated in agricultural wastewater treatment processes, whereas industrial wastewater is treated in industrial wastewater treatment processes.

For municipal wastewater the use of septic tanks and other On-Site Sewage Facilities (OSSF) is widespread in some rural areas, for example serving up to 20 percent of the homes in the U.S.

One type of aerobic treatment system is the activated sludge process, based on the maintenance and recirculation of a complex biomass composed of micro-organisms able to absorb and adsorb the organic matter carried in the wastewater. Anaerobic wastewater treatment processes (UASB, EGSB) are also widely applied in the treatment of industrial wastewaters and biological sludge. Some wastewater may be highly treated and reused as reclaimed water. Constructed wetlands are also being used.

Reuse

Treated wastewater can be reused as drinking water, in industry (for example in cooling towers), in artificial recharge of aquifers, in agriculture and in the rehabilitation of natural ecosystems (for example in Florida's Everglades).

Reuse in Gardening and Agriculture

There are benefits of using recycled water for irrigation, including the lower cost compared to some other sources and consistency of supply regardless of season, climatic conditions and associated water restrictions. Irrigation with recycled wastewater can also serve to fertilize plants if it contains nutrients, such as nitrogen, phosphorus and potassium.

Around 90% of wastewater produced globally remains untreated, causing widespread water pollution, especially in low-income countries. Increasingly, agriculture is using untreated wastewater for irrigation. Cities provide lucrative markets for fresh produce, so are attractive to farmers. However, because agriculture has to compete for increasingly scarce water resources with industry and municipal users, there is often no alternative for farmers but to use water polluted with urban waste directly to water their crops.

Health Risks of Polluted Irrigation Water

There can be significant health hazards related to using untreated wastewater in agriculture. Wastewater from cities can contain a mixture of chemical and biological pollutants.

In low-income countries, there are often high levels of pathogens from excreta, while in emerging nations, where industrial development is outpacing environmental regulation, there are increasing risks from inorganic and organic chemicals. The World Health Organization, in collaboration with the Food and Agriculture Organization of the United Nations (FAO) and the United Nations Environmental Program (UNEP), has developed guidelines for safe use of wastewater in 2006. These guidelines advocate a 'multiple-barrier' approach to wastewater use, for example by encouraging farmers to adopt various risk-reducing behaviours. These include ceasing irrigation a few days before harvesting to allow pathogens to die off in the sunlight, applying water carefully so it does not contaminate leaves likely to be eaten raw, cleaning vegetables with disinfectant or allowing fecal sludge used in farming to dry before being used as a human manure.

Etymology

The words "sewage" and "sewer" came from Old French *essouier* = "to drain", which came from Latin *exaquāre*. Their formal Latin antecedents are *exaquāticum* and *exaquārium*.

Legislation

European Union

Council Directive 91/271/EEC on Urban Wastewater Treatment was adopted on 21 May 1991, amended by the Commission Directive 98/15/EC. Commission Decision 93/481/EEC defines the information that Member States should provide the Commission on the state of implementation of the Directive.

United States

The Clean Water Act is the primary federal law in the United States governing water pollution.

Philippines

In the Philippines, Republic Act 9275, otherwise known as the Philippine Clean Water Act of 2004, is the governing law on wastewater management. It states that it is the country's policy to protect, preserve and revive the quality of our fresh, brackish and marine waters, for which wastewater management plays a particular role.

Suspended Solids

Suspended solids refers to small solid particles which remain in suspension in water as a colloid or due to the motion of the water. It is used as one indicator of water quality.

It is sometimes abbreviated **SS**, but is not to be confused with settleable solids, also abbreviated SS, which contribute to the blocking of sewer pipes.

Explanation

Suspended solids are important as pollutants and pathogens are carried on the surface of particles. The smaller the particle size, the greater the total surface area per unit mass of particle in grams, and so the higher the pollutant load that is likely to be carried.

Removal

Removal of suspended solids is generally achieved through the use of sedimentation and/or water filters (usually at a municipal level). By eliminating most of the suspended solids in a water supply, the significant water is usually rendered close to drinking quality. This is followed by disinfection to ensure that any free floating pathogens, or pathogens associated with the small remaining amount of suspended solids, are rendered ineffective.

Effectiveness of Filtering

The use of a very simple cloth filter, consisting of a folded cotton sari, drastically reduces the load of cholera carried in the water, and is suitable for use by the very poor; in this case, an appropriate technology method of disinfection might be added, such as solar water disinfection.

A major exception to this generalization is arsenic contamination of groundwater, as arsenic is a very serious pollutant which is soluble, and thus not removed when suspended solids are removed. This makes it very difficult to remove, and finding an alternative water source is often the most realistic option.

References

- Maczulak, Anne Elizabeth (2010). Pollution: Treating Environmental Toxins. New York: Infobase Publishing. p. 120. ISBN 9781438126333.

- "Toxic Waste Facts, information, pictures | Encyclopedia.com articles about Toxic Waste". www.encyclopedia.com. Retrieved 2016-05-07.

- Hallam, Bill (April–May 2010). "Techniques for Efficient Hazardous Chemicals Handling and Disposal". Pollution Equipment News. p. 13. Retrieved 10 March 2016.

- "LABORATORY CHEMICAL WASTE MANAGEMENT GUIDELINES" (PDF). Environmental Health and Radiation Safety University of Pennsylvania. Retrieved 10 March 2016.

- "Waste - Disposal of Laboratory Wastes (GUIDANCE) | Current Staff | University of St Andrews". www.st-andrews.ac.uk. Retrieved 2016-02-04.

- Laboratory, National Research Council (US) Committee on Prudent Practices in the (2011-01-01). "Management of Waste". Retrieved 10 March 2016.

- "PROCEDURES FOR LABORATORY CHEMICAL WASTE DISPOSAL" (PDF). Memorial University. Retrieved 10 March 2016.

- "An Act Providing For A Comprehensive Water Quality Management And For Other Purposes". The LawPhil Project. Retrieved September 30, 2016.

- Jiaranaikhajorn, Taweechai. "Waste and Hazardous Substances Management Bureau" (PDF). Pollution Control Department (PCD), THAILAND. Retrieved 22 November 2014.

- Visvanathan, C. "Hazardous and Industrial Solid Waste Management in Thailand - an Overview" (PDF). www.faculty.ait.ac.th/visu/. Asian Institute of Technology Thailand. Retrieved 22 November 2014.

- "Genco Background". General Environment Conservation Public Comapany Limited (GENCO). Retrieved 22 November 2014.

Industrial Wastewater Treatment

This chapter has been carefully written to provide an easy understanding of the varied aspects of industrial wastewater treatment. The processes and mechanisms used to treat wastewater expelled by industries comprise industrial wastewater treatments. Some of the techniques used for this are reverse osmosis, ion-exchange resin, flocculation and activated sludge.

Industrial Wastewater Treatment

Industrial wastewater treatment covers the mechanisms and processes used to treat wastewater that is produced as a by-product of industrial or commercial activities. After treatment, the treated industrial wastewater (or effluent) may be reused or released to a sanitary sewer or to a surface water in the environment. Most industries produce some wastewater although recent trends in the developed world have been to minimise such production or recycle such wastewater within the production process. However, many industries remain dependent on processes that produce wastewaters.

Sources of Industrial Wastewater

Complex Organic Chemicals Industry

A range of industries manufacture or use complex organic chemicals. These include pesticides, pharmaceuticals, paints and dyes, petrochemicals, detergents, plastics, paper pollution, etc. Waste waters can be contaminated by feedstock materials, by-products, product material in soluble or particulate form, washing and cleaning agents, solvents and added value products such as plasticisers. Treatment facilities that do not need control of their effluent typically opt for a type of aerobic treatment, i.e. aerated lagoons.

Electric Power Plants

Fossil-fuel power stations, particularly coal-fired plants, are a major source of industrial wastewater. Many of these plants discharge wastewater with significant levels of metals such as lead, mercury, cadmium and chromium, as well as arsenic, selenium and nitrogen compounds (nitrates and nitrites). Wastewater streams include flue-gas desulfurization, fly ash, bottom ash and flue gas mercury control. Plants with air pollution controls such as wet scrubbers typically transfer the captured pollutants to the wastewater stream.

Ash ponds, a type of surface impoundment, are a widely used treatment technology at coal-fired plants. These ponds use gravity to settle out large particulates (measured as total suspended solids) from power plant wastewater. This technology does not treat dissolved pollutants. Power stations use additional technologies to control pollutants, depending on the particular wastestream in the plant. These include dry ash handling, closed-loop ash recycling, chemical precipitation, biological treatment (such as an activated sludge process), and evaporation.

Food Industry

Wastewater generated from agricultural and food operations has distinctive characteristics that set it apart from common municipal wastewater managed by public or private sewage treatment plants throughout the world: it is biodegradable and non-toxic, but has high concentrations of biochemical oxygen demand (BOD) and suspended solids (SS). The constituents of food and agriculture wastewater are often complex to predict, due to the differences in BOD and pH in effluents from vegetable, fruit, and meat products and due to the seasonal nature of food processing and post-harvesting.

Processing of food from raw materials requires large volumes of high grade water. Vegetable washing generates waters with high loads of particulate matter and some dissolved organic matter. It may also contain surfactants.

Animal slaughter and processing produces very strong organic waste from body fluids, such as blood, and gut contents. This wastewater is frequently contaminated by significant levels of antibiotics and growth hormones from the animals and by a variety of pesticides used to control external parasites.

Processing food for sale produces wastes generated from cooking which are often rich in plant organic material and may also contain salt, flavourings, colouring material and acids or alkali. Very significant quantities of oil or fats may also be present.

Iron and Steel Industry

The production of iron from its ores involves powerful reduction reactions in blast furnaces. Cooling waters are inevitably contaminated with products especially ammonia and cyanide. Production of coke from coal in coking plants also requires water cooling and the use of water in by-products separation. Contamination of waste streams includes gasification products such as benzene, naphthalene, anthracene, cyanide, ammonia, phenols, cresols together with a range of more complex organic compounds known collectively as polycyclic aromatic hydrocarbons (PAH).

The conversion of iron or steel into sheet, wire or rods requires hot and cold mechanical transformation stages frequently employing water as a lubricant and coolant. Contaminants include hydraulic oils, tallow and particulate solids. Final treatment of iron and steel products before onward sale into manufacturing includes *pickling* in strong mineral acid to remove rust and prepare the surface for tin or chromium plating or for other surface treatments such as galvanisation or painting. The two acids commonly used are hydrochloric acid and sulfuric acid. Wastewaters include acidic rinse waters together with waste acid. Although many plants operate acid recovery plants (particularly those using hydrochloric acid), where the mineral acid is boiled away from the iron salts, there remains a large volume of highly acid ferrous sulfate or ferrous chloride to be disposed of. Many steel industry wastewaters are contaminated by hydraulic oil, also known as *soluble oil*.

Mines and Quarries

The principal waste-waters associated with mines and quarries are slurries of rock particles in water. These arise from rainfall washing exposed surfaces and haul roads and also from rock washing and grading processes. Volumes of water can be very high, especially rainfall related arisings on large sites. Some specialized separation operations, such as coal washing to separate coal from native rock using density gradients, can produce wastewater contaminated by fine particulate haematite and surfactants. Oils and hydraulic oils are also common contaminants.

Wastewater from metal mines and ore recovery plants are inevitably contaminated by the minerals present in the native rock formations. Following crushing and extraction of the desirable materials, undesirable materials may enter the wastewater stream. For metal mines, this can include unwanted metals such as zinc and other materials such as arsenic. Extraction of high value metals such as gold and silver may generate slimes containing very fine particles in where physical removal of contaminants becomes particularly difficult.

Additionally, the geologic formations that harbour economically valuable metals such as copper and gold very often consist of sulphide-type ores. The processing entails grinding the rock into fine particles and then extracting the desired metal(s), with the leftover rock being known as tailings. These tailings contain a combination of not only undesirable leftover metals, but also sulphide components which eventually form sulphuric acid upon the exposure to air and water that inevitably occurs when the tailings are disposed of in large impoundments. The resulting acid mine drainage, which is often rich in heavy metals (because acids dissolve metals), is one of the many environmental impacts of mining.

Nuclear Industry

The waste production from the nuclear and radio-chemicals industry is dealt with as *Radioactive waste*.

Pulp and Paper Industry

Effluent from the pulp and paper industry is generally high in suspended solids and BOD. Plants that bleach wood pulp for paper making may generate chloroform, dioxins (including 2,3,7,8-TCDD), furans, phenols and chemical oxygen demand (COD). Stand-alone paper mills using imported pulp may only require simple primary treatment, such as sedimentation or dissolved air flotation. Increased BOD or COD loadings, as well as organic pollutants, may require biological treatment such as activated sludge or upflow anaerobic sludge blanket reactors. For mills with high inorganic loadings like salt, tertiary treatments may be required, either general membrane treatments like ultrafiltration or reverse osmosis or treatments to remove specific contaminants, such as nutrients.

Industrial Oil Contamination

Industrial applications where oil enters the wastewater stream may include vehicle wash bays, workshops, fuel storage depots, transport hubs and power generation. Often the wastewater is discharged into local sewer or trade waste systems and must meet local environmental specifications. Typical contaminants can include solvents, detergents, grit. lubricants and hydrocarbons.

Water Treatment

Many industries have a need to treat water to obtain very high quality water for demanding purposes such as environmental discharge compliance. Water treatment produces organic and mineral sludges from filtration and sedimentation. Ion exchange using natural or synthetic resins removes calcium, magnesium and carbonate ions from water, typically replacing them with sodium, chloride, hydroxyl and/or other ions. Regeneration of ion exchange columns with strong acids and alkalis produces a wastewater rich in hardness ions which are readily precipitated out, especially when in admixture with other wastewater constituents.

Wool Processing

Insecticide residues in fleeces are a particular problem in treating waters generated in wool processing. Animal fats may be present in the wastewater, which if not contaminated, can be recovered for the production of tallow or further rendering.

Treatment of Industrial Wastewater

The various types of contamination of wastewater require a variety of strategies to remove the contamination.

Brine Treatment

Brine treatment involves removing dissolved salt ions from the waste stream. Although similarities to seawater or brackish water desalination exist, industrial brine treatment may contain unique combinations of dissolved ions, such as hardness ions or other metals, necessitating specific processes and equipment.

Brine treatment systems are typically optimized to either reduce the volume of the final discharge for more economic disposal (as disposal costs are often based on volume) or maximize the recovery of fresh water or salts. Brine treatment systems may also be optimized to reduce electricity consumption, chemical usage, or physical footprint.

Brine treatment is commonly encountered when treating cooling tower blowdown, produced water from steam assisted gravity drainage (SAGD), produced water from natural gas extraction such as coal seam gas, frac flowback water, acid mine or acid rock drainage, reverse osmosis reject, chlor-alkali wastewater, pulp and paper mill effluent, and waste streams from food and beverage processing.

Brine treatment technologies may include: membrane filtration processes, such as reverse osmosis; ion exchange processes such as electrodialysis or weak acid cation exchange; or evaporation processes, such as brine concentrators and crystallizers employing mechanical vapour recompression and steam.

Reverse osmosis may not be viable for brine treatment, due to the potential for fouling caused by hardness salts or organic contaminants, or damage to the reverse osmosis membranes from hydrocarbons.

Evaporation processes are the most widespread for brine treatment as they enable the highest degree of concentration, as high as solid salt. They also produce the highest purity effluent, even distillate-quality. Evaporation processes are also more tolerant of organics, hydrocarbons, or hardness salts. However, energy consumption is high and corrosion may be an issue as the prime mover is concentrated salt water. As a result, evaporation systems typically employ titanium or duplex stainless steel materials.

Brine Management

Brine management examines the broader context of brine treatment and may include consideration of government policy and regulations, corporate sustainability, environmental impact, recycling, handling and transport, containment, centralized compared to on-site treatment, avoidance and reduction, technologies, and economics. Brine management shares some issues with leachate management and more general waste management.

Solids Removal

Most solids can be removed using simple sedimentation techniques with the solids recovered as slurry or sludge. Very fine solids and solids with densities close to the density of water pose special problems. In such case filtration or ultrafiltration may be required. Although, flocculation may be used, using alum salts or the addition of polyelectrolytes.

Oils and Grease Removal

The effective removal of oils and grease is dependent on the characteristics of the oil in terms of its suspension state and droplet size, which will in turn affect the choice of separator technology.

Oil pollution in water usually comes in four states, often in combination:

- free oil - large oil droplets sitting on the surface;

- heavy oil, which sits at the bottom, often adhering to solids like dirt;

- emulsified, where the oil droplets are heavily "chopped"; and

- dissolved oil, where the droplets are fully dispersed and not visible. Emulsified oil droplets are the most common in industrial oily wastewater and are extremely difficult to separate.

The methodology for separating the oil is dependent on the oil droplet size. Larger oil droplets such as those in free oil pollution are easily removed, but as the droplets become smaller, some separator technologies perform better than others.

Most separator technologies will have an optimum range of oil droplet sizes that can be effectively treated. This is known as the "micron rating."

Analysing the oily water to determine droplet size can be performed with a video particle analyser. Alternatively, there are commonalities in industries for oil droplet sizes. Larger droplets–greater than 60 microns–are often present in wastewater in workshops, re-fuel areas and depots. Twenty to 50 micron oil droplets often are present in vehicle wash bays, meat processing and dairy manufacturing effluent and aluminium billet cooling towers. Smaller droplets in the range of 10 to 20 microns tend to occur in workshops and condensates.

Each separator technology will have its' own performance curve outlining optimum performance based on oil droplet size. the most common separators are gravity tanks or pits, API oil-water separators or plate packs, chemical treatment via DAFs, centrifuges, media filters and hydrocyclones.

API separators

Many oils can be recovered from open water surfaces by skimming devices. Considered a dependable and cheap way to remove oil, grease and other hydrocarbons from water, oil skimmers can sometimes achieve the desired level of water purity. At other times, skimming is also a cost-efficient method to remove most of the oil before using membrane filters and chemical processes. Skimmers will prevent filters from blinding prematurely and keep chemical costs down because there is less oil to process.

Because grease skimming involves higher viscosity hydrocarbons, skimmers must be equipped with heaters powerful enough to keep grease fluid for discharge. If floating grease forms into solid clumps or mats, a spray bar, aerator or mechanical apparatus can be used to facilitate removal.

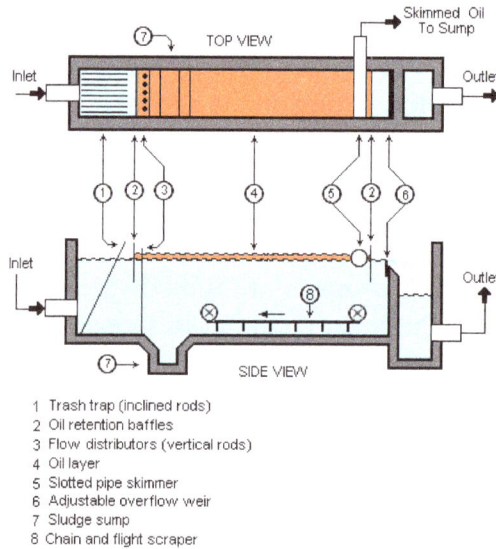

1 Trash trap (inclined rods)
2 Oil retention baffles
3 Flow distributors (vertical rods)
4 Oil layer
5 Slotted pipe skimmer
6 Adjustable overflow weir
7 Sludge sump
8 Chain and flight scraper

A typical API oil-water separator used in many industries

However, hydraulic oils and the majority of oils that have degraded to any extent will also have a soluble or emulsified component that will require further treatment to eliminate. Dissolving or emulsifying oil using surfactants or solvents usually exacerbates the problem rather than solving it, producing wastewater that is more difficult to treat.

The wastewaters from large-scale industries such as oil refineries, petrochemical plants, chemical plants, and natural gas processing plants commonly contain gross amounts of oil and suspended solids. Those industries use a device known as an API oil-water separator which is designed to separate the oil and suspended solids from their wastewater effluents. The name is derived from the fact that such separators are designed according to standards published by the American Petroleum Institute (API).

The API separator is a gravity separation device designed by using Stokes Law to define the rise velocity of oil droplets based on their density and size. The design is based on the specific gravity difference between the oil and the wastewater because that difference is much smaller than the specific gravity difference between the suspended solids and water. The suspended solids settles to the bottom of the separator as a sediment layer, the oil rises to top of the separator and the cleansed wastewater is the middle layer between the oil layer and the solids.

Typically, the oil layer is skimmed off and subsequently re-processed or disposed of, and the bottom sediment layer is removed by a chain and flight scraper (or similar device) and a sludge pump. The water layer is sent to further treatment for additional removal of any residual oil and then to some type of biological treatment unit for removal of undesirable dissolved chemical compounds.

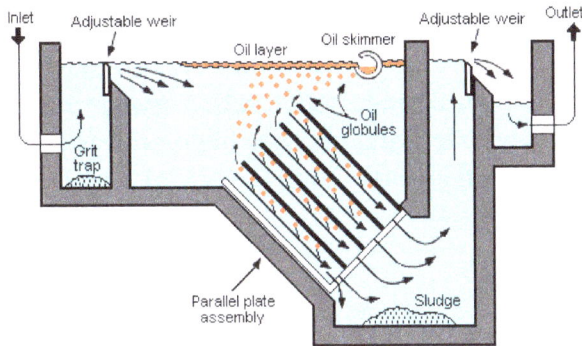

A typical parallel plate separator

Parallel plate separators are similar to API separators but they include tilted parallel plate assemblies (also known as parallel packs). The parallel plates provide more surface for suspended oil droplets to coalesce into larger globules. Such separators still depend upon the specific gravity between the suspended oil and the water. However, the parallel plates enhance the degree of oil-water separation. The result is that a parallel plate separator requires significantly less space than a conventional API separator to achieve the same degree of separation.

Hydrocyclone oil separators

Hydrocyclone oil separators operate on the process where wastewater enters the cyclone chamber and is spun under extreme centrifugal forces more than 1000 times the force of gravity. This force causes the water and oil droplets to separate. The separated oil is discharged from one end of the cyclone where treated water is discharged through the opposite end for further treatment, filtration or discharge.

Hydrocyclones are useful for the greatest range of oil droplet sizes operating from less than 10 microns and up and can operate continuously without water pre-treatment and at any temperature and pH. Applications where hydrocyclones are found are in industry where oily water sources arise in workshops, vehicle wash bays, transport hubs, fuel depots and aluminium billet processing. Animal fats from meat processing and dairy manufacturing can also be removed without the need of chemical treatment that often is required for dissolved air flotation (DAF) systems.

Removal of Biodegradable Organics

Biodegradable organic material of plant or animal origin is usually possible to treat using extended conventional sewage treatment processes such as activated sludge or trickling filter. Problems can arise if the wastewater is excessively diluted with washing water or is highly concentrated such as undiluted blood or milk. The presence of cleaning agents, disinfectants, pesticides, or antibiotics can have detrimental impacts on treatment processes.

Activated Sludge Process

Activated sludge is a biochemical process for treating sewage and industrial wastewater that uses air (or oxygen) and microorganisms to biologically oxidize organic pollutants, producing a waste sludge (or floc) containing the oxidized material. In general, an activated sludge process includes:

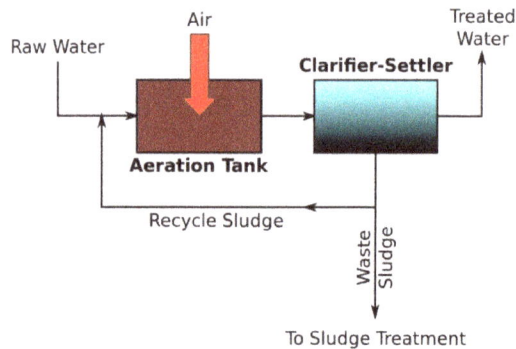

A generalized diagram of an activated sludge process.

- An aeration tank where air (or oxygen) is injected and thoroughly mixed into the wastewater.

- A settling tank (usually referred to as a clarifier or "settler") to allow the waste sludge to settle. Part of the waste sludge is recycled to the aeration tank and the remaining waste sludge is removed for further treatment and ultimate disposal.

Trickling Filter Process

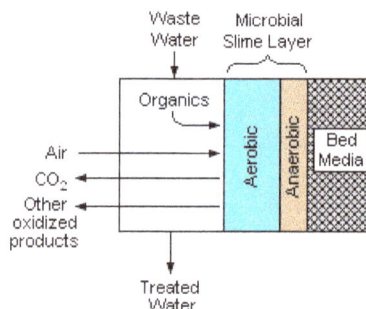

Image 1: A schematic cross-section of the contact face of the bed media in a trickling filter

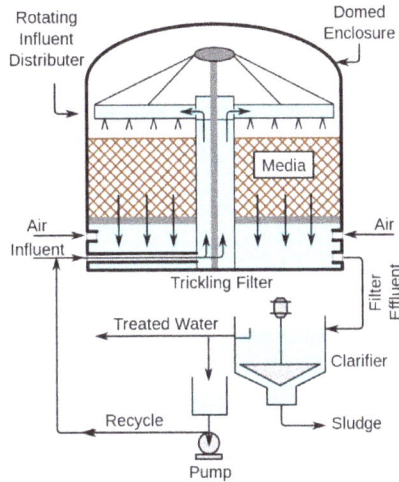

A typical complete trickling filter system

A trickling filter consists of a bed of rocks, gravel, slag, peat moss, or plastic media over which wastewater flows downward and contacts a layer (or film) of microbial slime covering the bed media. Aerobic conditions are maintained by forced air flowing through the bed or by natural convection of air. The process involves adsorption of organic compounds in the wastewater by the microbial slime layer, diffusion of air into the slime layer to provide the oxygen required for the biochemical oxidation of the organic compounds. The end products include carbon dioxide gas, water and other products of the oxidation. As the slime layer thickens, it becomes difficult for the air to penetrate the layer and an inner anaerobic layer is formed.

The fundamental components of a complete trickling filter system are:

- A bed of filter medium upon which a layer of microbial slime is promoted and developed.

- An enclosure or a container which houses the bed of filter medium.

- A system for distributing the flow of wastewater over the filter medium.

- A system for removing and disposing of any sludge from the treated effluent.

The treatment of sewage or other wastewater with trickling filters is among the oldest and most well characterized treatment technologies.

A trickling filter is also often called a *trickle filter*, *trickling biofilter*, *biofilter*, *biological filter* or *biological trickling filter*.

Treatment of Other Organics

Synthetic organic materials including solvents, paints, pharmaceuticals, pesticides, products from coke production and so forth can be very difficult to treat. Treatment methods are often specific to the material being treated. Methods include advanced oxidation processing, distillation, adsorption, vitrification, incineration, chemical immobilisation or landfill disposal. Some materials such as some detergents may be capable of biological degradation and in such cases, a modified form of wastewater treatment can be used.

Treatment of Acids and Alkalis

Acids and alkalis can usually be neutralised under controlled conditions. Neutralisation frequently produces a precipitate that will require treatment as a solid residue that may also be toxic. In some cases, gases may be evolved requiring treatment for the gas stream. Some other forms of treatment are usually required following neutralisation.

Waste streams rich in hardness ions as from de-ionisation processes can readily lose the hardness ions in a buildup of precipitated calcium and magnesium salts. This precipitation process can cause severe *furring* of pipes and can, in extreme cases, cause the blockage of disposal pipes. A 1 metre diameter industrial marine discharge pipe serving a major chemicals complex was blocked by such salts in the 1970s. Treatment is by concentration of de-ionisation waste waters and disposal to landfill or by careful pH management of the released wastewater.

Treatment of Toxic Materials

Toxic materials including many organic materials, metals (such as zinc, silver, cadmium, thallium, etc.) acids, alkalis, non-metallic elements (such as arsenic or selenium) are generally resistant to biological processes unless very dilute. Metals can often be precipitated out by changing the pH or by treatment with other chemicals. Many, however, are resistant to treatment or mitigation and may require concentration followed by landfilling or recycling. Dissolved organics can be *incinerated* within the wastewater by the advanced oxidation process.

Reverse Osmosis

Reverse osmosis (RO) is a water purification technology that uses a semipermeable membrane to remove ions, molecules, and larger particles from drinking water. In reverse osmosis, an applied pressure is used to overcome osmotic pressure, a colligative property, that is driven by chemical potential differences of the solvent, a thermodynamic parameter. Reverse osmosis can remove many types of dissolved and suspended species from water, including bacteria, and is used in both industrial processes and the production of potable water. The result is that the solute is retained on the pressurized side of the membrane and the pure solvent is allowed to pass to the other side. To be "selective", this membrane should not allow large molecules or ions through the pores (holes), but should allow smaller components of the solution (such as solvent molecules) to pass freely.

In the normal osmosis process, the solvent naturally moves from an area of low solute concentration (high water potential), through a membrane, to an area of high solute concentration (low water potential). The driving force for the movement of the solvent is the reduction in the free energy of the system when the difference in solvent concentration on either side of a membrane is reduced, generating osmotic pressure due to the solvent moving into the more concentrated solution. Applying an external pressure to reverse the natural flow of pure solvent, thus, is reverse osmosis. The process is similar to other membrane technology applications. However, key differences are found between reverse osmosis and filtration. The predominant removal mechanism in membrane filtration is straining, or size exclusion, so the process can theoretically achieve perfect efficiency regardless of parameters such as the solution's pressure and concentration. Reverse osmosis also involves diffusion, making the process dependent on pressure, flow rate, and other conditions. Reverse osmosis is most commonly known for its use in drinking water purification from seawater, removing the salt and other effluent materials from the water molecules.

History

The process of osmosis through semipermeable membranes was first observed in 1748 by Jean-Antoine Nollet. For the following 200 years, osmosis was only a phenomenon observed in the laboratory. In 1950, the University of California at Los Angeles first investigated desalination of seawater using semipermeable membranes. Researchers from both University of California at Los Angeles and the University of Florida successfully produced fresh water from seawater in the mid-1950s, but the flux was too low to be commercially viable until the discovery at University of California at Los Angeles by Sidney Loeb and Srinivasa Sourirajan at the National Research Council of Canada, Ottawa, of techniques for making asymmetric membranes characterized by an effectively thin "skin" layer supported atop a highly porous and much thicker substrate region of the membrane. John Cadotte, of FilmTec Corporation, discovered that membranes with particularly high flux and low salt passage could be made by interfacial polymerization of m-phenylene diamine and trimesoyl chloride. Cadotte's patent on this process was the subject of litigation and has since expired. Almost all commercial reverse osmosis membrane is now made by this method. By the end of 2001, about 15,200 desalination plants were in operation or in the planning stages, worldwide.

In 1977 Cape Coral, Florida became the first municipality in the United States to use the RO process on a large scale with an initial operating capacity of 3 million gallons (11350 m^3) per day. By 1985, due to the rapid growth in population of Cape Coral, the city had the largest low pressure reverse osmosis plant in the world, capable of producing 15 million gallons per day (MGD) (56800 m^3/d).

Reverse osmosis production train, North Cape Coral Reverse Osmosis Plant

Process

Osmosis is a natural process. When two solutions with different concentrations of a solute are separated by a semipermeable membrane, the solvent has a tendency to move from low to high solute concentrations for chemical potential equilibration.

A semipermeable membrane coil used in desalination

Formally, reverse osmosis is the process of forcing a solvent from a region of high solute concentration through a semipermeable membrane to a region of low solute concentration by applying a pressure in excess of the osmotic pressure. The largest and most important application of reverse osmosis is the separation of pure water from seawater and brackish waters; seawater or brackish water is pressurized against one surface of the membrane, causing transport of salt-depleted water across the membrane and emergence of potable drinking water from the low-pressure side.

The membranes used for reverse osmosis have a dense layer in the polymer matrix—either the skin of an asymmetric membrane or an interfacially polymerized layer within a thin-film-composite membrane—where the separation occurs. In most cases, the membrane is designed to allow only water to pass through this dense layer, while preventing the passage of solutes (such as salt ions). This process requires that a high pressure be exerted on the high concentration side of the membrane, usually 2–17 bar (30–250 psi) for fresh and brackish water, and 40–82 bar (600–1200 psi) for seawater, which has around 27 bar (390 psi) natural osmotic pressure that must be overcome. This process is best known for its use in desalination (removing the salt and other minerals from sea water to get fresh water), but since the early 1970s, it has also been used to purify fresh water for medical, industrial, and domestic applications.

Fresh Water Applications

Drinking Water Purification

Around the world, household drinking water purification systems, including a reverse osmosis step, are commonly used for improving water for drinking and cooking.

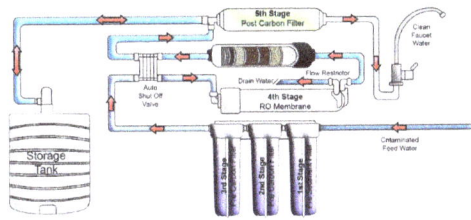

The reverse osmosis water filter process

Such systems typically include a number of steps:

- a sediment filter to trap particles, including rust and calcium carbonate

- optionally, a second sediment filter with smaller pores

- an activated carbon filter to trap organic chemicals and chlorine, which will attack and degrade thin film composite membrane reverse osmosis membranes

- a reverse osmosis filter, which is a thin film composite membrane

- optionally, a second carbon filter to capture those chemicals not removed by the reverse osmosis membrane

- optionally an ultraviolet lamp for sterilizing any microbes that may escape filtering by the reverse osmosis membrane

- latest developments in the sphere include nano materials and membranes

In some systems, the carbon prefilter is omitted, and cellulose triacetate membrane is used. CTA (Cellulose Triacetate) membrane is a paper by-product membrane bonded to a synthetic layer and are made to allow contact with chlorine in the water. These require a small amount of chlorine in the water source to prevent bacteria from forming on it. The typical rejection rate for CTA membranes is 85-95%.

The cellulose triacetate membrane is prone to rotting unless protected by chlorinated water, while the thin film composite membrane is prone to breaking down under the influence of chlorine. A thin film composite (TFC) membranes are made of synthetic material, and require chlorine to be removed before the water enters the membrane. To protect the TFC membrane elements from chlorine damage, carbon filters are used as pre-treatment in all residential reverse osmosis systems. TFC membranes have a higher rejection rate of 95-98% and a longer life than CTA membranes.

Portable reverse osmosis water processors are sold for personal water purification in various locations. To work effectively, the water feeding to these units should be under some pressure (40 pounds per square inch (280 kPa) or greater is the norm). Portable reverse osmosis water processors can be used by people who live in rural areas without clean water, far away from the city's water pipes. Rural people filter river or ocean water themselves, as the device is easy to use (saline water may need special membranes). Some travelers on long boating, fishing, or island camping trips, or in countries where the local water supply is polluted or substandard, use reverse osmosis water processors coupled with one or more ultraviolet sterilizers.

In the production of bottled mineral water, the water passes through a reverse osmosis water processor to remove pollutants and microorganisms. In European countries, though, such processing of natural mineral water (as defined by a European Directive) is not allowed under European law. In practice, a fraction of the living bacteria can and do pass through reverse osmosis membranes through minor imperfections, or bypass the membrane entirely through tiny leaks in surrounding seals. Thus, complete reverse osmosis systems may include additional water treatment stages that use ultraviolet light or ozone to prevent microbiological contamination.

Membrane pore sizes can vary from 0.1 to 5,000 nm (4×10^{-9} to 2×10^{-4} in) depending on filter type. Particle filtration removes particles of 1 µm (3.9×10^{-5} in) or larger. Microfiltration removes particles of 50 nm or larger. Ultrafiltration removes particles of roughly 3 nm or larger. Nanofiltration removes particles of 1 nm or larger. Reverse osmosis is in the final category of membrane filtration, hyperfiltration, and removes particles larger than 0.1 nm.

Military Use: The Reverse Osmosis Water Purification Unit

United States Marines from Combat Logistics Battalion 31 operate reverse osmosis water purification units for relief efforts after the 2006 Southern Leyte mudslide

A reverse osmosis water purification unit (ROWPU) is a portable, self-contained water treatment plant. Designed for military use, it can provide potable water from nearly any water source. There are many models in use by the United States armed forces and the Canadian Forces. Some models are containerized, some are trailers, and some are vehicles unto themselves.

Each branch of the United States armed forces has their own series of reverse osmosis water purification unit models, but they are all similar. The water is pumped from its raw source into the reverse osmosis water purification unit module, where it is treated with a polymer to initiate coagulation. Next, it is run through a multi-media filter where it undergoes primary treatment by removing turbidity. It is then pumped through a cartridge filter which is usually spiral-wound cotton. This process clarifies the water of any particles larger than 5 micrometres (0.00020 in) and eliminates almost all turbidity.

The clarified water is then fed through a high-pressure piston pump into a series of vessels where it is subject to reverse osmosis. The product water is free of 90.00–99.98% of the raw water's total dissolved solids and by military standards, should have no more than 1000–1500 parts per million by measure of electrical conductivity. It is then disinfected with chlorine and stored for later use.

Within the United States Marine Corps, the reverse osmosis water purification unit has been replaced by both the Lightweight Water Purification System and Tactical Water Purification Systems. The Lightweight Water Purification Systems can be transported by Humvee and filters 125 US gallons (470 l) per hour. The Tactical Water Purification Systems can be carried on a Medium Tactical Vehicle Replacement truck, and can filter 1,200 to 1,500 US gallons (4,500 to 5,700 l) per hour.

Water and Wastewater Purification

Rain water collected from storm drains is purified with reverse osmosis water processors and used for landscape irrigation and industrial cooling in Los Angeles and other cities, as a solution to the problem of water shortages.

In industry, reverse osmosis removes minerals from boiler water at power plants. The water is distilled multiple times. It must be as pure as possible so it does not leave deposits on the machinery or cause corrosion. The deposits inside or outside the boiler tubes may result in underperformance of the boiler, bringing down its efficiency and resulting in poor steam production, hence poor power production at the turbine.

It is also used to clean effluent and brackish groundwater. The effluent in larger volumes (more than 500 m³/d) should be treated in an effluent treatment plant first, and then the clear effluent is subjected to reverse osmosis system. Treatment cost is reduced significantly and membrane life of the reverse osmosis system is increased.

The process of reverse osmosis can be used for the production of deionized water.

Reverse osmosis process for water purification does not require thermal energy. Flow-through reverse osmosis systems can be regulated by high-pressure pumps. The recovery of purified water depends upon various factors, including membrane sizes, membrane pore size, temperature, operating pressure, and membrane surface area.

In 2002, Singapore announced that a process named NEWater would be a significant part of its future water plans. It involves using reverse osmosis to treat domestic wastewater before discharging the NEWater back into the reservoirs.

Food Industry

In addition to desalination, reverse osmosis is a more economical operation for concentrating food liquids (such as fruit juices) than conventional heat-treatment processes. Research has been done on concentration of orange juice and tomato juice. Its advantages include a lower operating cost and the ability to avoid heat-treatment processes, which makes it suitable for heat-sensitive substances such as the protein and enzymes found in most food products.

Reverse osmosis is extensively used in the dairy industry for the production of whey protein powders and for the concentration of milk to reduce shipping costs. In whey applications, the whey (liquid remaining after cheese manufacture) is concentrated with reverse osmosis from 6% total solids to 10–20% total solids before ultrafiltration processing. The ultrafiltration retentate can then be used to make various whey powders, including whey protein isolate. Additionally, the ultrafiltration permeate, which contains lactose, is concentrated by reverse osmosis from 5% total solids to 18–22% total solids to reduce crystallization and drying costs of the lactose powder.

Although use of the process was once avoided in the wine industry, it is now widely understood and used. An estimated 60 reverse osmosis machines were in use in Bordeaux, France, in 2002. Known users include many of the elite classed growths (Kramer) such as Château Léoville-Las Cases in Bordeaux.

Maple Syrup Production

In 1946, some maple syrup producers started using reverse osmosis to remove water from sap before the sap is boiled down to syrup. The use of reverse osmosis allows about 75–90% of the water to be removed from the sap, reducing energy consumption and exposure of the syrup to high temperatures. Microbial contamination and degradation of the membranes must be monitored.

Hydrogen Production

For small-scale hydrogen production, reverse osmosis is sometimes used to prevent formation of minerals on the surface of electrodes.

Reef Aquariums

Many reef aquarium keepers use reverse osmosis systems for their artificial mixture of seawater. Ordinary tap water can contain excessive chlorine, chloramines, copper, nitrates, nitrites, phosphates, silicates, or many other chemicals detrimental to the sensitive organisms in a reef environment. Contaminants such as nitrogen compounds and phosphates can lead to excessive and unwanted algae growth. An effective combination of both reverse osmosis and deionization is the most popular among reef aquarium keepers, and is preferred above other water purification processes due to the low cost of ownership and minimal operating costs. Where chlorine and chloramines are found in the water, carbon filtration is needed before the membrane, as the common residential membrane used by reef keepers does not cope with these compounds.

Window Cleaning

An increasingly popular method of cleaning windows is the so-called "water-fed pole" system. Instead of washing the windows with detergent in the conventional way, they are scrubbed with highly purified water, typically containing less than 10 ppm dissolved solids, using a brush on the end of a long pole which is wielded from ground level. Reverse osmosis is commonly used to purify the water.

Landfill Leachate Purification

Treatment with reverse osmosis is limited, resulting in low recoveries on high concentration (measured with electrical conductivity) and fouling of the RO membranes. Reverse osmosis applicability is limited by conductivity, organics, and scaling inorganic elements such as $CaSO_4$, Si, Fe and Ba. Low organic scaling can be used two different technology, one is using spiral wound membrane type of module, and for high organic scaling, high conductivity and higher pressure (up to 90 bars) can be used disc tube module with reverse osmosis membranes. Disc tube modules was redesigned for landfill leachate purification, what usually is contaminated with high organics. Due to the cross-flow with high velocity is given by a flow booster pump, what is recirculating the flow over the same membrane surface between 1,5 and 3 times before is released as a concentrate. High velocity is also good against membrane scaling and successful membrane cleanings.

Power Consumption for A Disc Tube Module System

disc tube module   spiral wound module

Disc tube module with RO membrane cushion and Spiral wound module with RO membrane

energy consumption per m³ leachate			
name of module	1-stage up to 75 bar	2-stage up to 75 bar	3-stage up to 120 bar
disc tube module	6.1 – 8.1 kWh/m³	8.1 – 9.8 kWh/m³	11.2 – 14.3 kWh/m³

Desalination

Areas that have either no or limited surface water or groundwater may choose to desalinate. Reverse osmosis is an increasingly common method of desalination, because of its relatively low energy consumption. In recent years, energy consumption has dropped to around 3 kWh/m³, with the development of more efficient energy recovery devices and improved membrane materials. According to the International Desalination Association, for 2011, reverse osmosis was used in 66% of installed desalination capacity (44.5 of 67.4 Mm³/day), and nearly all new plants. Other plants mainly use thermal distillation methods: multiple-effect distillation and multi-stage flash.

Sea water reverse osmosis (SWRO) desalination, a membrane process, has been commercially used since the early 1970s. Its first practical use was demonstrated by Sidney Loeb from University of California at Los Angeles in Coalinga, California, and Srinivasa Sourirajan of National Research council, Canada. Because no heating or phase changes are needed, energy requirements are low, around 3 kWh/m³, in comparison to other processes of desalination, but are still much higher than those required for other forms of water supply, including reverse osmosis treatment of wastewater, at 0.1 to 1 kWh/m³. Up to 50% of the seawater input can be recovered as fresh water, though lower recoveries may reduce membrane fouling and energy consumption.

Brackish water reverse osmosis refers to desalination of water with a lower salt content than sea water, usually from river estuaries or saline wells. The process is substantially the same as sea water reverse osmosis, but requires lower pressures and therefore less energy. Up to 80% of the feed water input can be recovered as fresh water, depending on feed salinity.

The Ashkelon sea water reverse osmosis desalination plant in Israel is the largest in the world. The project was developed as a build-operate-transfer by a consortium of three international companies: Veolia water, IDE Technologies, and Elran.

The typical single-pass sea water reverse osmosis system consists of:

- Intake

- Pretreatment

- High pressure pump (if not combined with energy recovery)

- Membrane assembly

- Energy recovery (if used)

- Remineralisation and pH adjustment

- Disinfection

- Alarm/control panel

Pretreatment

Pretreatment is important when working with reverse osmosis and nanofiltration membranes due to the nature of their spiral-wound design. The material is engineered in such a fashion as to allow only one-way flow through the system. As such, the spiral-wound design does not allow for backpulsing with water or air agitation to scour its surface and remove solids. Since accumulated material cannot be removed from the membrane surface systems, they are highly susceptible to fouling (loss of production capacity). Therefore, pretreatment is a necessity for any reverse osmosis or nanofiltration system. Pretreatment in sea water reverse osmosis systems has four major components:

- Screening of solids: Solids within the water must be removed and the water treated to prevent fouling of the membranes by fine particle or biological growth, and reduce the risk of damage to high-pressure pump components.

- Cartridge filtration: Generally, string-wound polypropylene filters are used to remove particles of 1–5 μm diameter.

- Dosing: Oxidizing biocides, such as chlorine, are added to kill bacteria, followed by bisulfite dosing to deactivate the chlorine, which can destroy a thin-film composite membrane. There are also biofouling inhibitors, which do not kill bacteria, but simply prevent them from growing slime on the membrane surface and plant walls.

- Prefiltration pH adjustment: If the pH, hardness and the alkalinity in the feedwater result in a scaling tendency when they are concentrated in the reject stream, acid is dosed to maintain carbonates in their soluble carbonic acid form.

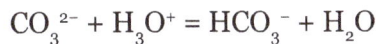

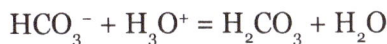

$$CO_3^{2-} + H_3O^+ = HCO_3^- + H_2O$$

$$HCO_3^- + H_3O^+ = H_2CO_3 + H_2O$$

- Carbonic acid cannot combine with calcium to form calcium carbonate scale. Calcium carbonate scaling tendency is estimated using the Langelier saturation index. Adding too much sulfuric acid to control carbonate scales may result in calcium sulfate, barium sulfate, or strontium sulfate scale formation on the reverse osmosis membrane.

- Prefiltration antiscalants: Scale inhibitors (also known as antiscalants) prevent formation of all scales compared to acid, which can only prevent formation of calcium carbonate and calcium phosphate scales. In addition to inhibiting carbonate and phosphate scales, antiscalants inhibit sulfate and fluoride scales and disperse colloids and metal oxides. Despite claims that antiscalants can inhibit silica formation, no concrete evidence proves that silica polymerization can be inhibited by antiscalants. Antiscalants can control acid-soluble scales at a fraction of the dosage required to control the same scale using sulfuric acid.

- Some small scale desalination units use 'beach wells'; they are usually drilled on the seashore in close vicinity to the ocean. These intake facilities are relatively simple to build and the seawater they collect is pretreated via slow filtration through the subsurface sand/seabed formations in the area of source water extraction. Raw seawater collected using beach wells is often of better quality in terms of solids, silt, oil and grease, natural organic contamination and aquatic microorganisms, compared to open seawater intakes. Sometimes, beach intakes may also yield source water of lower salinity.

High Pressure Pump

The high pressure pump supplies the pressure needed to push water through the membrane, even as the membrane rejects the passage of salt through it. Typical pressures for brackish water range from 225 to 376 psi (15.5 to 26 bar, or 1.6 to 2.6 MPa). In the case of seawater, they range from 800 to 1,180 psi (55 to 81.5 bar or 6 to 8 MPa). This requires a large amount of energy. Where energy recovery is used, part of the high pressure pump's work is done by the energy recovery device, reducing the system energy inputs.

Membrane Assembly

The membrane assembly consists of a pressure vessel with a membrane that allows feedwater to be pressed against it. The membrane must be strong enough to withstand whatever pressure is applied against it. Reverse osmosis membranes are made in a variety of configurations, with the two most common configurations being spiral-wound and hollow-fiber.

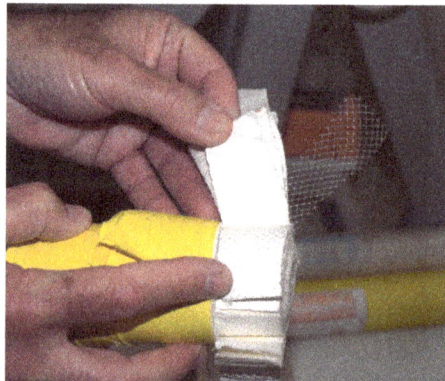

The layers of a membrane

Only a part of the saline feed water pumped into the membrane assembly passes through the membrane with the salt removed. The remaining "concentrate" flow passes along the saline side of the membrane to flush away the concentrated salt solution. The percentage of desalinated water produced versus the saline water feed flow is known as the "recovery ratio". This varies with the salinity of the feed water and the system design parameters: typically 20% for small seawater systems, 40% – 50% for larger seawater systems, and 80% – 85% for brackish water. The concentrate flow is at typically only 3 bar / 50 psi less than the feed pressure, and thus still carries much of the high pressure pump input energy.

The desalinated water purity is a function of the feed water salinity, membrane selection and recovery ratio. To achieve higher purity a second pass can be added which generally requires re-pumping. Purity expressed as total dissolved solids typically varies from 100 to 400 parts per million (ppm or milligram/litre)on a seawater feed. A level of 500 ppm is generally accepted as the upper limit for drinking water, while the US Food and Drug Administration classifies mineral water as water containing at least 250 ppm.

Energy Recovery

Energy recovery can reduce energy consumption by 50% or more. Much of the high pressure pump input energy can be recovered from the concentrate flow, and the increasing efficiency of energy recovery devices has greatly reduced the energy needs of reverse osmosis desalination. Devices used, in order of invention, are:

- Turbine or Pelton wheel: a water turbine driven by the concentrate flow, connected to the high pressure pump drive shaft to provide part of its input power. Positive displacement axial piston motors have also been used in place of turbines on smaller systems.

- Turbocharger: a water turbine driven by the concentrate flow, directly connected to a centrifugal pump which boosts the high pressure pump output pressure, reducing the pressure needed from the high pressure pump and thereby its energy input, similar in construction principle to car engine turbochargers.

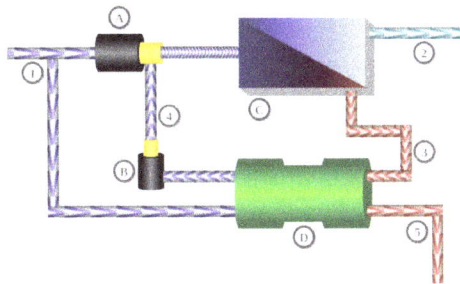

Schematics of a reverse osmosis desalination system using a pressure exchanger.
1: Sea water inflow,
2: Fresh water flow (40%),
3: Concentrate flow (60%),
4: Sea water flow (60%),

5: Concentrate (drain),
A: Pump flow (40%),
B: Circulation pump,
C: Osmosis unit with membrane,
D: Pressure exchanger

- Pressure exchanger: using the pressurized concentrate flow, in direct contact or via a piston, to pressurize part of the membrane feed flow to near concentrate flow pressure. A boost pump then raises this pressure by typically 3 bar / 50 psi to the membrane feed pressure. This reduces flow needed from the high-pressure pump by an amount equal to the concentrate flow, typically 60%, and thereby its energy input. These are widely used on larger low-energy systems. They are capable of 3 kWh/m³ or less energy consumption.

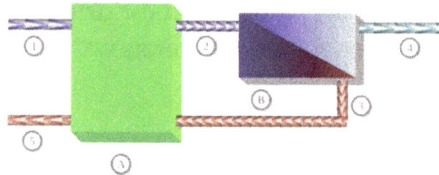

Schematic of a reverse osmosis desalination system using an energy recovery pump.
1: Sea water inflow (100%, 1 bar),
2: Sea water flow (100%, 50 bar),
3: Concentrate flow (60%, 48 bar),
4: Fresh water flow (40%, 1 bar),
5: Concentrate to drain (60%,1 bar),
A: Pressure recovery pump,
B: Osmosis unit with membrane

- Energy recovery pump: a reciprocating piston pump having the pressurized concentrate flow applied to one side of each piston to help drive the membrane feed flow from the opposite side. These are the simplest energy recovery devices to apply, combining the high pressure pump and energy recovery in a single self-regulating unit. These are widely used on smaller low-energy systems. They are capable of 3 kWh/m³ or less energy consumption.

Remineralisation and pH Adjustment

The desalinated water is "stabilized" to protect downstream pipelines and storage, usually by adding lime or caustic soda to prevent corrosion of concrete-lined surfaces. Liming material is used to adjust pH between 6.8 and 8.1 to meet the potable water specifications, primarily for effective disinfection and for corrosion control. Remineralisation may be needed to replace minerals removed from the water by desalination. Although this process has proved to be costly and not very convenient if it is intended to meet mineral demand by humans and plants. The very same mineral demand that freshwater sources provided previously. For instance water from Israel's national water carrier typically contains dissolved magnesium levels of 20 to 25 mg/liter, while water from the Ashkelon plant has no magnesium. After farmers used this water, magnesium deficiency symptoms appeared in crops, including tomatoes, basil, and flowers, and had to be remedied by fertilization. Current Israeli drinking water standards set a minimum calcium level of 20 mg/liter. The postdesalination treatment in the Ashkelon plant uses sulfuric acid to dissolve calcite (limestone), resulting in calcium concentration of 40 to 46 mg/liter. This is still lower than the 45 to 60 mg/liter found in typical Israeli freshwaters.

Disinfection

Post-treatment consists of preparing the water for distribution after filtration. Reverse osmosis is an effective barrier to pathogens, but post-treatment provides secondary protection against compromised membranes and downstream problems. Disinfection by means of ultra violet (UV) lamps (sometimes called germicidal or bactericidal) may be employed to sterilize pathogens which bypassed the reverse osmosis process. Chlorination or chloramination (chlorine and ammonia) protects against pathogens which may have lodged in the distribution system downstream, such as from new construction, backwash, compromised pipes, etc.

Disadvantages

Household reverse osmosis units use a lot of water because they have low back pressure. As a result, they recover only 5 to 15% of the water entering the system. The remainder is discharged as waste water. Because waste water carries with it the rejected contaminants, methods to recover this water are not practical for household systems. Wastewater is typically connected to the house drains and will add to the load on the household septic system. A reverse osmosis unit delivering five gallons (19 L) of treated water per day may discharge between 20 and 90 gallons (75-340 L) of waste water per day.

Large-scale industrial/municipal systems recover typically 75% to 80% of the feed water, or as high as 90%, because they can generate the high pressure needed for higher recovery reverse osmosis filtration. On the other hand, as recovery of wastewater increases in commercial operations, effective contaminant removal rates tend to become reduced, as evidenced by product water total dissolved solids levels.

Due to its fine membrane construction, reverse osmosis not only removes harmful contaminants present in the water, but it also may remove many of the desirable minerals from the water. A number of peer-reviewed studies have looked at the long-term health effects of drinking demineralized water.

Waste Stream Considerations

Depending upon the desired product, either the solvent or solute stream of reverse osmosis will be waste. For food concentration applications, the concentrated solute stream is the product and the solvent stream is waste. For water treatment applications, the solvent stream is purified water and the solute stream is concentrated waste. The solvent waste stream from food processing may be used as reclaimed water, but there may be fewer options for disposal of a concentrated waste solute stream. Ships may use marine dumping and coastal desalination plants typically use marine outfalls. Landlocked reverse osmosis plants may require evaporation ponds or injection wells to avoid polluting groundwater or surface runoff.

New Developments

Since the 1970s, prefiltration of high-fouling waters with another larger-pore membrane, with less hydraulic energy requirement, has been evaluated and sometimes used. However, this means that the water passes through two membranes and is often repressurized, which requires more energy to be put into the system, and thus increases the cost.

Other recent developmental work has focused on integrating reverse osmosis with electrodialysis to improve recovery of valuable deionized products, or to minimize the volume of concentrate requiring discharge or disposal.

In the production of drinking water, the latest developments include nanoscale and graphene membranes.

The world's largest RO desalination plant was built in Sorek, Israel in 2013. It has an output of 624,000 m^3/day. It is also the cheapest and will sell water to the authorities for US$0.58/$m^3$.

Electrodialysis

Electrodialysis (ED) is used to transport salt ions from one solution through ion-exchange membranes to another solution under the influence of an applied electric potential difference. This is done in a configuration called an electrodialysis cell. The cell consists of a feed (dilute) compartment and a concentrate (brine) compartment formed by an anion exchange membrane and a cation exchange membrane placed between two electrodes. In almost all practical electrodialysis processes, multiple electrodialysis cells are arranged into a configuration called an electrodialysis stack, with alternating anion and cation exchange membranes forming the multiple electrodialysis cells. Electrodialysis processes are different from distillation techniques and other membrane based processes (such as reverse osmosis (RO)) in that dissolved species are moved away from the feed stream rather than the reverse. Because the quantity of dissolved species in the feed stream is far less than that of the fluid, electrodialysis offers the practical advantage of much higher feed recovery in many applications.

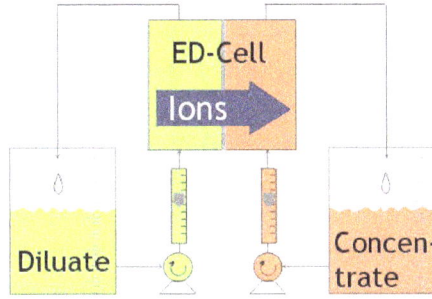

Method

In an electrodialysis stack, the dilute (D) feed stream, brine or concentrate (C) stream, and electrode (E) stream are allowed to flow through the appropriate cell compartments formed by the ion exchange membranes. Under the influence of an electrical potential difference, the negatively charged ions (e.g., chloride) in the dilute stream migrate toward the positively charged anode. These ions pass through the positively charged anion exchange membrane, but are prevented from further migration toward the anode by the negatively charged cation exchange membrane and therefore stay in the C stream, which becomes concentrated with the anions. The positively charged species (e.g., sodium) in the D stream migrate toward the negatively charged cathode and pass through the negatively charged cation exchange membrane. These cations also stay in the C stream, prevented from further migration toward the cathode by the positively charged anion exchange membrane. As a result of the anion and cation migration, electric current flows between the cathode and anode. Only an equal number of anion and cation charge equivalents are transferred from the D stream into the C stream and so the charge balance is maintained in each stream. The overall result of the electrodialysis process is an ion concentration increase in the concentrate stream with a depletion of ions in the dilute solution feed stream.

The E stream is the electrode stream that flows past each electrode in the stack. This stream may consist of the same composition as the feed stream (e.g., sodium chloride) or may be a separate solution containing a different species (e.g., sodium sulfate). Depending on the stack configuration, anions and cations from the electrode stream may be transported into the C stream, or anions and cations from the D stream may be transported into the E stream. In each case, this transport is necessary to carry current across the stack and maintain electrically neutral stack solutions.

Anode and Cathode Reactions

Reactions take place at each electrode. At the cathode,

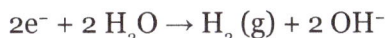

$$2e^- + 2\,H_2O \rightarrow H_2\,(g) + 2\,OH^-$$

while at the anode,

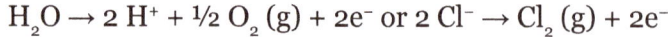

$$H_2O \rightarrow 2\,H^+ + \tfrac{1}{2}\,O_2\,(g) + 2e^-\ \text{or}\ 2\,Cl^- \rightarrow Cl_2\,(g) + 2e^-$$

Small amounts of hydrogen gas are generated at the cathode and small amounts of either oxygen or chlorine gas (depending on composition of the E stream and end ion exchange membrane arrangement) at the anode. These gases are typically subsequently dissipated as the E stream effluent from each electrode compartment is combined to maintain a neutral pH and discharged or re-circulated to a separate E tank. However, some (e.g.,) have proposed collection of hydrogen gas for use in energy production.

Efficiency

Current efficiency is a measure of how effective ions are transported across the ion exchange membranes for a given applied current. Typically current efficiencies >80% are desirable in commercial stacks to minimize energy operating costs. Low current efficiencies indicate water splitting in the diluate or concentrate streams, shunt currents between the electrodes, or back-diffusion of ions from the concentrate to the diluate could be occurring.

Current efficiency is calculated according to:

$$\xi = \frac{zFQ_f\left(C_{inlet}^d - C_{outlet}^d\right)}{NI}$$

where

ξ = current utilization efficiency

z = charge of the ion

F = Faraday constant, 96,485 Amp-s/mol

Q_f = dilute flow rate, L/s

C_{inlet}^d = dilute ED cell inlet concentration, mol/L

C_{outlet}^d = dilute ED cell outlet concentration, mol/L

N = number of cell pairs

I = current, Amps.

Current efficiency is generally a function of feed concentration.

Applications

In application, electrodialysis systems can be operated as continuous production or batch production processes. In a continuous process, feed is passed through a sufficient number of stacks placed in series to produce the final desired product quality. In batch processes, the diluate and/or concentrate streams are re-circulated through the electrodialysis systems until the final product or concentrate quality is achieved.

Electrodialysis is usually applied to deionization of aqueous solutions. However, desalting of sparingly conductive aqueous organic and organic solutions is also possible. Some applications of electrodialysis include:

- Large scale brackish and seawater desalination and salt production.

- Small and medium scale drinking water production (e.g., towns & villages, construction & military camps, nitrate reduction, hotels & hospitals)

- Water reuse (e.g., industrial laundry wastewater, produced water from oil/gas production, cooling tower makeup & blowdown, metals industry fluids, wash-rack water)

- Pre-demineralization (e.g., boiler makeup & pretreatment, ultrapure water pretreatment, process water desalination, power generation, semiconductor, chemical manufacturing, food and beverage)

- Food processing

- Agricultural water (e.g., water for greenhouses, hydroponics, irrigation, live-stock)

- Glycol desalting (e.g., antifreeze / engine-coolants, capacitor electrolyte fluids, oil and gas dehydration, conditioning and processing solutions, industrial heat transfer fluids, secondary coolants from heating, venting, and air conditioning (HVAC))

- Glycerin purification

The major application of electrodialysis has historically been the desalination of brackish water or seawater as an alternative to RO for potable water production and seawater concentration for salt production (primarily in Japan). In normal potable water production without the requirement of high recoveries, reverse osmosis is generally believed to be more cost-effective when total dissolved solids (TDS) are 3,000 parts per million (ppm) or greater, while electrodialysis is more cost-effective for TDS feed concentrations less than 3,000 ppm or when high recoveries of the feed are required.

Another important application for electrodialysis is the production of pure water and ultrapure water by electrodeionization (EDI). In EDI, the purifying compartments and sometimes the concentrating compartments of the electrodialysis stack are filled with ion exchange resin. When fed with low TDS feed (e.g., feed purified by RO), the product can reach very high purity levels (e.g., 18 MΩ-cm). The ion exchange resins act to retain the ions, allowing these to be transported across the ion exchange membranes. The main usage of EDI systems are in electronics, pharmaceutical, power generation, and cooling tower applications.

Limitations

Electrodialysis has inherent limitations, working best at removing low molecular weight ionic components from a feed stream. Non-charged, higher molecular weight, and less mobile ionic species will not typically be significantly removed. Also, in contrast to RO, electrodialysis becomes less economical when extremely low salt concentrations in the product are required and with sparingly conductive feeds: current density becomes limited and current utilization efficiency typically decreases as the feed salt concentration becomes lower, and with fewer ions in solution to carry current, both ion transport and energy efficiency greatly declines. Consequently, comparatively large membrane areas are required to satisfy capacity requirements for low concentration (and sparingly conductive) feed solutions. Innovative systems overcoming the inherent limitations of electrodialysis (and RO) are available; these integrated systems work synergistically, with each sub-system operating in its optimal range, providing the least overall operating and capital costs for a particular application.

As with RO, electrodialysis systems require feed pretreatment to remove species that coat, precipitate onto, or otherwise "foul" the surface of the ion exchange membranes. This fouling decreases the efficiency of the electrodialysis system. Species of concern include calcium and magnesium hardness, suspended solids, silica, and organic compounds. Water softening can be used to remove hardness, and micrometre or multimedia filtration can be used to remove suspended solids. Hardness in particular is a concern since scaling can build up on the membranes. Various chemicals are also available to help prevent scaling. Also, electrodialysis reversal systems seek to minimize scaling by periodically reversing the flows of diluate and concentrate and polarity of the electrodes.

Ion-Exchange Resin

An ion-exchange resin or ion-exchange polymer is a resin or polymer that acts as a medium for ion exchange. It is an insoluble matrix (or support structure) normally in the form of small (0.5–1 mm diameter) microbeads, usually white or yellowish, fabricated from an organic polymer substrate. The beads are typically porous, providing a large surface area on and inside them. The trapping of ions occurs along with the accompanying release of other ions, and thus the process is called ion exchange. There are multiple types of ion-exchange resin. Most commercial resins are made of polystyrene sulfonate.

Ion-exchange resin beads

Ion-exchange resins are widely used in different separation, purification, and decontamination processes. The most common examples are water softening and water purification. In many cases ion-exchange resins were introduced in such processes as a more flexible alternative to the use of natural or artificial zeolites. Also, ion-exchange resins are highly effective in the biodiesel filtration process.

Ion-exchange resin beads

Types of Resins

Most typical ion-exchange resins are based on crosslinked polystyrene. The actual ion-exchanging sites are introduced after polymerisation. Additionally, in the case of polystyrene, crosslinking is introduced by copolymerisation of styrene and a few percent of divinylbenzene (non-crosslinked polymers are soluble in water). Crosslinking decreases ion-exchange capacity of the resin and prolongs the time needed to accomplish the ion-exchange processes but improves the robustness of the resin. Particle size also influences the resin parameters; smaller particles have larger outer surface, but cause larger head loss in the column processes.

Besides being made as bead-shaped materials, ion-exchange resins are also produced as membranes. These ion-exchange membranes, which are made of highly cross-linked ion-exchange resins that allow passage of ions, but not of water, are used for electrodialysis.

Four main types of ion-exchange resins differ in their functional groups:

- strongly acidic, typically featuring sulfonic acid groups, e.g. sodium polystyrene sulfonate or polyAMPS,

- strongly basic, typically featuring quaternary amino groups, for example, trimethylammonium groups, e.g. polyAPTAC),

- weakly acidic, typically featuring carboxylic acid groups,

- weakly basic, typically featuring primary, secondary, and/or tertiary amino groups, e.g. polyethylene amine.

Specialised ion-exchange resins are also known such as chelating resins (iminodiacetic acid, thiourea-based resins, and many others).

Anion resins and cation resins are the two most common resins used in the ion-exchange process. While anion resins attract negatively charged ions, cation resins attract positively charged ions.

Anion Resins

Anion resins may be either strongly or weakly basic. Strongly basic anion resins maintain their positive charge across a wide pH range, whereas weakly basic anion resins are neutralized at higher pH levels. Weakly basic resins do not maintain their charge at a high pH because they undergo deprotonation. They do, however, offer excellent mechanical and chemical stability. This, combined with a high rate of ion exchange, make weakly base anion resins well suited for the organic salts.

For anion resins, regeneration typically involves treatment of the resin with a strongly basic solution, e.g. aqueous sodium hydroxide. During regeneration, the regenerant chemical is passed through the resin, and trapped negative ions are flushed out, renewing the resin exchange capacity.

Cation-exchange Resin

Formula: R–H

Reaction:

$$2(R)-H + CaCl_2 = Ca(R_2) + 2HCl$$

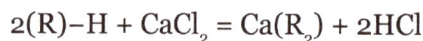

Anion-exchange Resin

Formula: $N(R)_4-OH$

Reaction:

$$N(R)_4-OH + HCl = N(R)_4Cl + H_2O$$

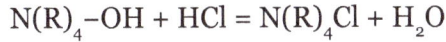

Anion-exchange chromatography makes use of this principle to extract and purify materials from mixtures or solutions.

Uses

Water Softening

In this application, ion-exchange resins are used to replace the magnesium and calcium ions found in hard water with sodium ions. When the resin is fresh, it contains sodium ions at its active sites. When in contact with a solution containing magnesium and calcium ions (but a low concentration of sodium ions), the magnesium and calcium ions preferentially migrate out of solution to the active sites on the resin, being replaced in solution by sodium ions. This process reaches equilibrium with a much lower concentration of magnesium and calcium ions in solution than was started with.

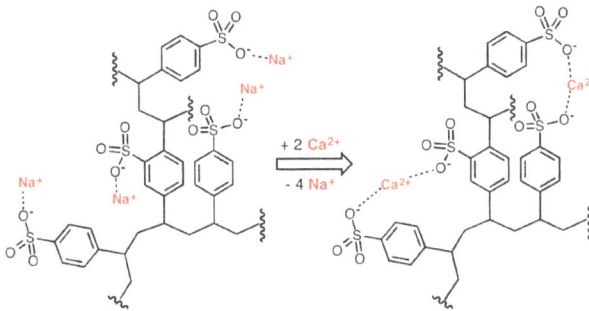

Idealised image of water softening process, involving replacement of calcium ions in water with sodium ions donated by a cation exchange resin.

The resin can be recharged by washing it with a solution containing a high concentration of sodium ions (e.g. it has large amounts of common salt (NaCl) dissolved in it). The calcium and magnesium ions migrate off the resin, being replaced by sodium ions from the solution until a new equilibrium is reached. The salt is used to recharge an ion-exchange resin which itself is used to soften the water.

Water Purification

In this application, ion-exchange resins are used to remove poisonous (e.g. copper) and heavy metal (e.g. lead or cadmium) ions from solution, replacing them with more innocuous ions, such as sodium and potassium.

Few ion-exchange resins remove chlorine or organic contaminants from water – this is usually done by using an activated charcoal filter mixed in with the resin. There are some ion-exchange resins that do remove organic ions, such as MIEX (magnetic ion-exchange) resins. Domestic water purification resin is not usually recharged – the resin is discarded when it can no longer be used.

Production of High-purity Water

Water of highest purity is required for electronics, scientific experiments, production of superconductors, and nuclear industry, among others. Such water is produced using ion-exchange processes or combinations of membrane and ion-exchange methods. Cations are replaced with hydrogen ions using cation-exchange resins; anions are replaced with hydroxyls using anion-exchange resins. The hydrogen ions and hydroxyls recombine producing water molecules. Thus, no ions remain in the produced water. The purification process is usually performed in several steps with "mixed bed ion-exchange columns" at the end of the technological chain.

Ion-exchange in Metal Separation

Ion-exchange processes are used to separate and purify metals, including separating uranium from plutonium and other actinides, including thorium; and lanthanum, neodymium, ytterbium, samarium, lutetium, from each other and the other lanthanides. There are two series of rare earth metals, the lanthanides and the actinides. Members of each family have very similar chemical and physical properties. Ion-exchange was for many years the only practical way to separate the rare earths in large quantities. This application was developed in the 1940s by Frank Spedding. Subsequently, solvent extraction has mostly supplanted use of ion exchange resins except for the highest purity products.

A drum of yellowcake

A very important case is the PUREX process (plutonium-uranium extraction process) which is used to separate the plutonium and the uranium from the spent fuel products from a nuclear reactor, and to be able to dispose of the waste products. Then, the plutonium and uranium are available for making nuclear-energy materials, such as new reactor fuel and nuclear weapons.

Ion-exchange beads are also an essential component in In-situ leach uranium mining. In-situ recovery involves the extraction of uranium-bearing water (grading as low as .05% U_3O_8) through boreholes. The extracted uranium solution is then filtered through the resin beads. Through an ion exchange process, the resin beads attract uranium from the solution. Uranium loaded resins are then transported to a processing plant, where U_3O_8 is separated from the resin beads and yellowcake is produced. The resin beads can then be returned to the ion exchange facility where they are reused.

The ion-exchange process is also used to separate other sets of very similar chemical elements, such as zirconium and hafnium, which incidentally is also very important for the nuclear industry. Zirconium is practically transparent to free neutrons, used in building reactors, but hafnium is a very strong absorber of neutrons, used in reactor control rods.

Catalysis

In chemistry ion-exchange resins in the acid form are known to catalyze organic reactions.

Juice Purification

Ion-exchange resins are used in the manufacture of fruit juices such as orange and cranberry juice, where they are used to remove bitter-tasting components and so improve the flavor. This allows tart or poorer-tasting fruit sources to be used for juice production.

Sugar Manufacturing

Ion-exchange resins are used in the manufacturing of sugar from various sources. They are used to help convert one type of sugar into another type of sugar, and to decolorize and purify sugar syrups.

Pharmaceuticals

Ion-exchange resins are used in the manufacturing of pharmaceuticals, not only for catalyzing certain reactions but also for isolating and purifying pharmaceutical active ingredients. Three ion-exchange resins, sodium polystyrene sulfonate, colestipol, and cholestyramine, are used as active ingredients. Sodium polystyrene sulfonate is a strongly acidic ion-exchange resin and is used to treat hyperkalemia. Colestipol is a weakly basic ion-exchange resin and is used to treat hypercholesterolemia. Cholestyramine is a strongly basic ion-exchange resin and is also used to treat hypercholesterolemia. Colestipol and cholestyramine are known as bile acid sequestrants.

Ion-exchange resins are also used as excipients in pharmaceutical formulations such as tablets, capsules, gums, and suspensions. In these uses the ion-exchange resin can have several different functions, including taste-masking, extended release, tablet dis-integration, increased bioavailability, and improving the chemical stability of the active ingredients.

Ultrafiltration

Ultrafiltration (UF) is a variety of membrane filtration in which forces like pressure or concentration gradients lead to a separation through a semipermeable membrane. Suspended solids and solutes of high molecular weight are retained in the so-called retentate, while water and low molecular weight solutes pass through the membrane in the permeate (filtrate). This separation process is used in industry and research for purifying and concentrating macromolecular (10^3 - 10^6 Da) solutions, especially protein solutions. Ultrafiltration is not fundamentally different from microfiltration. Both of these separate based on size exclusion or particle capture. It is fundamentally differ-ent from membrane gas separation, which separate based on different amounts of ab-sorption and different rates of diffusion. Ultrafiltration membranes are defined by the molecular weight cut-off (MWCO) of the membrane used. Ultrafiltration is applied in cross-flow or dead-end mode.

Applications

Industries such as chemical and pharmaceutical manufacturing, food and beverage processing, and waste water treatment, employ ultrafiltration in order to recycle flow or add value to later products. Blood dialysis also utilizes ultrafiltration.

Drinking Water

Drinking water treatment 300 m³/h using ultrafiltration in Grundmühle waterworks (Germany)

UF can be used for the removal of particulates and macromolecules from raw water to produce potable water. It has been used to either replace existing secondary (coagulation, flocculation, sedimentation) and tertiary filtration (sand filtration and chlorination) systems employed in water treatment plants or as standalone systems in isolated regions with growing populations. When treating water with high suspended solids, UF is often integrated into the process, utilising primary (screening, flotation, filtration) and some secondary treatments as pre-treatment stages. UF processes are currently preferred over traditional treatment methods for the following reasons:

- No chemicals required (aside from cleaning)

- Constant product quality regardless of feed quality

- Compact plant size

- Capable of exceeding regulatory standards of water quality, achieving 90-100% pathogen removal

UF processes are currently limited by the high cost incurred due to membrane fouling and replacement. Additional pretreatment of feed water is required to prevent excessive damage to the membrane units.

In many cases UF is used for pre filtration in reverse osmosis (RO) plants to protect the RO membranes.

Protein Concentration

UF is used extensively in the dairy industry; particularly in the processing of cheese whey to obtain whey protein concentrate (WPC) and lactose-rich permeate. In a single stage, a UF process is able to concentrate the whey 10-30 times the feed. The original alternative to membrane filtration of whey was using steam heating followed by drum drying or spray drying. The product of these methods had limited applications due to its granulated texture and insolubility. Existing methods also had inconsistent product composition, high capital and operating costs and due to the excessive heat used in drying would often denature some of the proteins. Compared to traditional methods, UF processes used for this application:

- Are more energy efficient

- Have consistent product quality, 35-80% protein product depending on operating conditions

- Do not denature proteins as they use moderate operating conditions

The potential for fouling is widely discussed, being identified as a significant contributor to decline in productivity. Cheese whey contains high concentrations of calcium phosphate which can potentially lead to scale deposits on the membrane surface. As a result substantial pretreatment must be implemented to balance pH and temperature of the feed to maintain solubility of calcium salts.

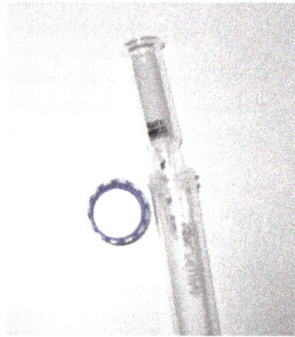

A selectively permeable membrane can be mounted in a centrifuge tube. The buffer is forced through the membrane by centrifugation, leaving the protein in the upper chamber.

Other Applications

- Filtration of effluent from paper pulp mill

- Cheese manufacture

- Removal of pathogens from milk

- Process and waste water treatment

- Enzyme recovery

- Fruit juice concentration and clarification

- Dialysis and other blood treatments

- Desalting and solvent-exchange of proteins (via diafiltration)

- Laboratory grade manufacturing

Principles

The basic operating principle of ultrafiltration uses a pressure induced separation of solutes from a solvent through a semi permeable membrane. The relationship between the applied pressure on the solution to be separated and the flux through the membrane is most commonly described by the Darcy equation:

$$J = \frac{TMP}{\mu R_t}$$

where J is the flux (flow rate per membrane area),TMP is the transmembrane pressure (pressure difference between feed and permeate stream), μ is solvent viscosity, R_t is the total resistance (sum of membrane and fouling resistance).

Membrane Fouling

Concentration Polarization

When filtration occurs the local concentration of rejected material at the membrane surface increases and can become saturated. In UF, increased ion concentration can develop an osmotic pressure on the feed side of the membrane. This reduces the effective TMP of the system, therefore reducing permeation rate. The increase in concentrated layer at the membrane wall decreases the permeate flux, due to increase in resistance which reduces the driving force for solvent to transport through membrane surface. CP affects almost all the available membrane separation process. In RO, the solutes retained at the membrane layer results in higher osmotic pressure in comparison to the bulk stream concentration. So the higher pressures are required to overcome this osmotic pressure. Concentration polarisation plays a dominant role in ultrafiltration as compared to microfiltration because of the small pore size membrane. It must be noted that concentration polarization differs from fouling as it has no lasting effects on the membrane itself and can be reversed by relieving the TMP. It does however have a significant effect on many types of fouling.

Types of Fouling

Particulate Deposition

The following models describe the mechanisms of particulate deposition on the membrane surface and in the pores:

- *Standard blocking*: macromolecules are uniformly deposited on pore walls

- *Complete blocking*: membrane pore is completely sealed by a macromolecule

- *Cake filtration*: accumulated particles or macromolecules form a fouling layer on the membrane surface, in UF this is also known as a gel layer

- *Intermediate blocking*: when macromolecules deposit into pores or onto already blocked pores, contributing to cake formation

Scaling

As a result of concentration polarization at the membrane surface, increased ion concentrations may exceed solubility thresholds and precipitate on the membrane surface. These inorganic salt deposits can block pores causing flux decline, membrane degradation and loss of production. The formation of scale is highly dependent on factors affecting both solubility and concentration polarization including pH, temperature, flow velocity and permeation rate.

Biofouling

Microorganisms will adhere to the membrane surface forming a gel layer – known as biofilm. The film increases the resistance to flow, acting as an additional barrier to permeation. In spiral-wound modules, blockages formed by biofilm can lead to uneven flow distribution and thus increase the effects of concentration polarization.

Membrane Arrangements

Depending on the shape and material of the membrane, different modules can be used for ultrafiltration process. Commercially available designs in ultrafiltration modules vary according to the required hydrodynamic and economic constraints as well as the mechanical stability of the system under particular operating pressures. The main modules used in industry include:

Hollow fibre module

Tubular Modules

The tubular module design uses polymeric membranes cast on the inside of plastic or porous paper components with diameters typically in the range of 5 – 25 mm with lengths from 0.6 - 6.4 m. Multiple tubes are housed in a PVC or steel shell. The feed of the module is passed through the tubes, accommodating radial transfer of permeate to the shell side. This design allows for easy cleaning however the main drawback is its low permeability, high volume hold-up within the membrane and low packing density.

Hollow Fibre

This design is conceptually similar to the tubular module with a shell and tube arrangement. A single module can consist of 50 to thousands of hollow fibres and therefore are self-supporting unlike the tubular design. The diameter of each fibre ranges from 0.2 – 3 mm with the feed flowing in the tube and the product permeate collected radially on the outside. The advantage of having self-supporting membranes as is the ease at which it can be cleaned due to its ability to be backflushed. Replacement costs however are high, as one faulty fibre will require the whole bundle to be replaced. Considering the tubes are of small diameter, using this design also makes the system prone to blockage.

Spiral-wound Modules

Are composed of a combination of flat membrane sheets separated by a thin meshed spacer material which serves as a porous plastic screen support. These sheets are rolled around a central perforated tube and fitted into a tubular steel pressure vessel casing. The feed solution passes over the membrane surface and the permeate spirals into the central collection tube. Spiral-wound modules are a compact and cheap alternative in ultrafiltration design, offer a high volumetric throughput and can also be easily cleaned. However it is limited by the thin channels where feed solutions with suspended solids can result in partial blockage of the membrane pores.

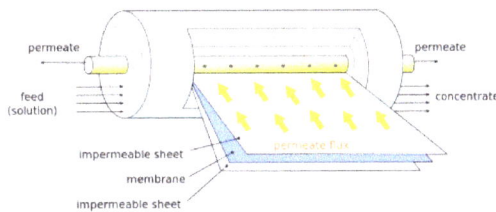

Spiral-wound membrane module

Plate and Frame

This uses a membrane placed on a flat plate separated by a mesh like material. The feed is passed through the system from which permeate is separated and collected from the edge of the plate. Channel length can range from 10 – 60 cm and channel heights from 0.5 – 1 mm. This module provides low volume hold-up, relatively easy replacement of the membrane and the ability to feed viscous solutions because of the low channel height, unique to this particular design.

Process Characteristics

The process characteristics of a UF system are highly dependent on the type of membrane used and its application. Manufacturers' specifications of the membrane tend to limit the process to the following typical specifications:

	Hollow Fibre	Spiral-wound	Ceramic Tubular	Plate and Frame
pH	2-13	2-11	3-7	
Feed Pressure (psi)	9-15	<30-120	60-100	

Backwash Pressure (psi)	9-15	20-40	10-30	
Temperature (°C)	5-30	5-45	5-400	
Total Dissolved Solids (mg/L)	<1000	<600	<500	
Total Suspended Solids (mg/L)	<500	<450	<300	
Turbidity (NTU)	<15	<1	<10	
Iron (mg/L)	<5	<5	<5	
Oils and Greases (mg/L)	<0.1	<0.1	<0.1	
Solvents, phenols (mg/L)	<0.1	<0.1	<0.1	

Process Design Considerations

When designing a new membrane separation facility or considering its integration into an existing plant, there are many factors which must be considered. For most applications a heuristic approach can be applied to determine many of these characteristics to simplify the design process. Some design areas include:

Pre-treatment

Treatment of feed prior to the membrane is essential to prevent damage to the membrane and minimize the effects of fouling which greatly reduce the efficiency of the separation. Types of pre-treatment are often dependent on the type of feed and its quality. For example in wastewater treatment, household waste and other particulates are screened. Other types of pre-treatment common to many UF processes include pH balancing and coagulation. Appropriate sequencing of each pre-treatment phase is crucial in preventing damage to subsequent stages. Pre-treatment can even be employed simply using dosing points.

Membrane Specifications

Material

Most UF membranes use polymer materials (polysulfone, polypropylene, cellulose acetate, polylactic acid) however ceramic membranes are used for high temperature applications.

Pore Size

A general rule for choice of pore size in a UF system is to use a membrane with a pore size one tenth that of the particle size to be separated. This limits the number of smaller particles entering the pores and adsorbing to the pore surface. Instead they block the entrance to the pores allowing simple adjustments of cross-flow velocity to dislodge them.

Operation Strategy

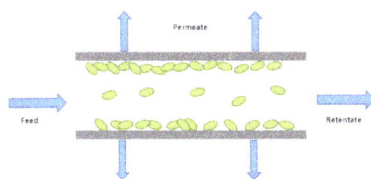

Schematic of cross flow operation.

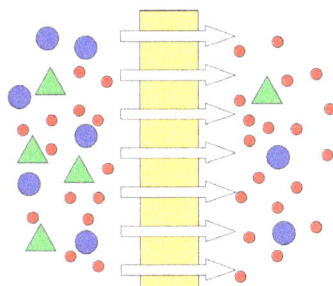

Schematic of dead-end operation

Flowtype

UF systems can either operate with cross-flow or dead-end flow. In dead-end filtration the flow of the feed solution is perpendicular to the membrane surface. On the other hand in cross flow systems the flow passes parallel to the membrane surface. Dead-end configurations are more suited to batch processes with low suspended solids as solids accumulate at the membrane surface therefore requiring frequent backflushes and cleaning to maintain high flux. Cross-flow configurations are preferred in continuous operations as solids are continuously flushed from the membrane surface resulting in a thinner cake layer and lower resistance to permeation.

Flow Velocity

Flow velocity is especially critical for hard water or liquids containing suspensions in preventing excessive fouling. Higher cross-flow velocities can be used to enhance the sweeping effect across the membrane surface therefore preventing deposition of macromolecules and colloidal material and reducing the effects of concentration polarization. Expensive pumps are however required to achieve these conditions.

Flow Temperature

To avoid excessive damage to the membrane, it is recommended to operate a plant at the temperature specified by the membrane manufacturer. In some instances however temperatures beyond the recommended region are required to minimise the effects of fouling. Economic analysis of the process is required to find a compromise between the increased cost of membrane replacement and productivity of the separation.

Pressure

Pressure drops over multi-stage separation can result in a drastic decline in flux performance in the latter stages of the process. This can be improved using booster pumps to increase the TMP in the final stages. This will incur a greater capital and energy cost which will be offset by the improved productivity of the process. With a multi-stage operation, retentate streams from each stage are recycled through the previous stage to improve their separation efficiency.

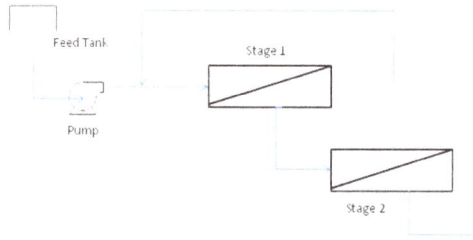

Typical two stage membrane process with recycle stream

Multi-Stage, Multi-module

Multiple stages in series can be applied to achieve higher purity permeate streams. Due to the modular nature of membrane processes, multiple modules can be arranged in parallel to treat greater volumes.

Post-treatment

Post-treatment of the product streams is dependent on the composition of the permeate and retentate and its end-use or government regulation. In cases such as milk separation both streams (milk and whey) can be collected and made into useful products. Additional drying of the retentate will produce whey powder. In the paper mill industry, the retentate (non-biodegradable organic material) is incinerated to recover energy and permeate (purified water) is discharged into waterways. It is essential for the permeate water to be pH balanced and cooled to avoid thermal pollution of waterways and altering its pH.

Cleaning

Cleaning of the membrane is done regularly to prevent the accumulation of foulants and reverse the degrading effects of fouling on permeability and selectivity. Regular backwashing is often conducted every 10 min for some processes to remove cake layers formed on the membrane surface. By pressurising the permeate stream and forcing it back through the membrane, accumulated particles can be dislodged, improving the flux of the process. Backwashing is limited in its ability to remove more complex forms of fouling such as biofouling, scaling or adsorption to pore walls. These types of foulants require chemical cleaning to be removed. The common types of chemicals used for cleaning are:

- Acidic solutions for the control of inorganic scale deposits

- Alkali solutions for removal of organic compounds

- Biocides or disinfection such as Chlorine or peroxide when bio-fouling is evident

When designing a cleaning protocol it is essential to consider:Cleaning time – Adequate time must be allowed for chemicals to interact with foulants and permeate into the membrane pores. However if the process is extended beyond its optimum duration it can lead to denaturation of the membrane and deposition of removed foulants. The complete cleaning cycle including rinses between stages may take as long as 2 hours to complete.Aggressiveness of chemical treatment – With a high degree of fouling it may be necessary to employ aggressive cleaning solutions to remove fouling material. However in some applications this may not be suitable if the membrane material is sensitive, leading to enhanced membrane ageing.Disposal of cleaning effluent – The release of some chemicals into wastewater systems may be prohibited or regulated therefore this must be considered. For example the use of phosphoric acid may result in high levels of phosphates entering water ways and must be monitored and controlled to prevent eutrophication.

Summary of Common Types of Fouling and Their Respective Chemical Treatments

Foulant	Reagent	Time and Temperature	Mode of Action
Fats and oils, proteins, polysaccharides, bacteria	0.5M NaOH with 200 ppm Cl2	30-60 min 25-55 °C	Hydrolysis and oxidation
DNA, mineral salts	0.1M – 0.5M acid (acetic, citric, nitric)	30-60 min 25-35 °C	Solubilization
Fats, oils, biopolymers, proteins	0.1% SDS, 0.1% Triton X-100	30 min – overnight 25-55 °C	Wetting, emulsifying, suspending, dispersing
Cell fragments, fats, oils, proteins	Enzyme detergents	30 min – overnight 30 – 40 °C	Catalytic breakdown
DNA	0.5% DNAase	30 min – overnight 20 – 40 °C	Enzyme hydrolysis

New Developments

In order to increase the life-cycle of membrane filtration systems, energy efficient membranes are being developed in membrane bioreactor systems. Technology has been introduced which allows the power required to aerate the membrane for cleaning to be reduced whilst still maintaining a high flux level. Mechanical cleaning processes have also been adopted using granulates as an alternative to conventional forms of cleaning; this reduces energy consumption and also reduces the area required for filtration tanks.

Membrane properties have also been enhanced to reduce fouling tendencies by modifying surface properties. This can be noted in the biotechnology industry where membrane surfaces have been altered in order to reduce the amount of protein binding. Ultrafiltration modules have also been improved to allow for more membrane for a given area without increasing its risk of fouling by designing more efficient module internals.

The current pre-treatment of seawater desulphonation uses ultrafiltration modules that have been designed to withstand high temperatures and pressures whilst occupying a smaller footprint. Each module vessel is self supported and resistant to corrosion and accommodates easy removal and replacement of the module without the cost of replacing the vessel itself.

Flocculation

Flocculation, in the field of chemistry, is a process wherein colloids come out of suspension in the form of floc or flake; either spontaneously or due to the addition of a clarifying agent. The action differs from precipitation in that, prior to flocculation, colloids are merely suspended in a liquid and not actually dissolved in a solution. In the flocculated system, there is no formation of a cake, since all the flocs are in the suspension.

Coagulation and flocculation are important processes in water treatment with coagulation to destabilize particles through chemical reaction between coagulant and colloids, and flocculation to transport the destabilized particles that will cause collisions with floc.

Term Definition

According to the IUPAC definition, flocculation is "a process of contact and adhesion whereby the particles of a dispersion form larger-size clusters." Flocculation is synonymous with agglomeration and coagulation / coalescence.

Basically, coagulation is a process of addition coagulant to destabilize a stabilized charged particle. Meanwhile, flocculation is a mixing technique that promotes agglomeration and assists in the settling of particles. The most common used coagulant is alum, $Al_2(SO_4)_3 \cdot 14H_2O$.

The chemical reaction involved:

$$Al_2(SO_4)_3 \cdot 14H_2O \rightarrow 2Al(OH)_3(s) + 6H^+ + 3SO_4^{2-} + 8H_2O$$

During flocculation, gentle mixing accelerates the rate of particle collision, and the destabilized particles are further aggregated and enmeshed into larger precipitates. Flocculation is affected by several parameters, including mixing speeds, mixing intensity, and mixing time. The product of the mixing intensity and mixing time is used to describe flocculation processes.

Applications

Surface Chemistry

In colloid chemistry, flocculation refers to the process by which fine particulates are caused to clump together into a floc. The floc may then float to the top of the liquid (creaming), settle to the bottom of the liquid (sedimentation), or be readily filtered from the liquid. Flocculation behavior of soil colloids is closely related to freshwater quality. High dispersibility of soil colloids not only directly causes turbidity of the surrounding water but it also induces eutrophication due to the adsorption of nutritional substances in rivers and lakes.

Physical Chemistry

For emulsions, flocculation describes clustering of individual dispersed droplets together, whereby the individual droplets do not lose their identity. Flocculation is thus the initial step leading to further ageing of the emulsion (droplet coalescence and the ultimate separation of the phases). Flocculation is used in mineral dressing.

Civil Engineering/Earth Sciences

In civil engineering, and in the earth sciences, flocculation is a condition in which clays, polymers or other small charged particles become attached and form a fragile structure, a floc. In dispersed clay slurries, flocculation occurs after mechanical agitation ceases and the dispersed clay platelets spontaneously form flocs because of attractions between negative face charges and positive edge charges.

Biology

Flocculation is used in biotechnology applications in conjunction with microfiltration to improve the efficiency of biological feeds. The addition of synthetic flocculants to the bioreactor can increase the average particle size making microfiltration more efficient. When flocculants are not added, cakes form and accumulate causing low cell viability. Positively charged flocculants work better than negatively charged ones since the cells are generally negatively charged.

Cheese Industry

Flocculation is widely employed to measure the progress of curd formation in the initial stages of cheese making to determine how long the curds must set. The reaction involving the rennet micelles are modeled by Smoluchowski kinetics. During the renneting of milk the micelles can approach one another and flocculate, a process that involves hydrolysis of molecules and macropeptides.

Flocculation is also used during cheese wastewater treatment. Three different coagulants are mainly used:

- $FeSO_4$ (Iron(II) sulfate)

- $Al_2(SO_4)_3$ (Aluminium sulfate)

- $FeCl_3$ (Iron(III) chloride)

Brewing

In the brewing industry flocculation has a different meaning. It is a very important process in fermentation during the production of beer where cells form macroscopic flocs. These flocs cause the yeast to sediment at the end of the fermentation. Subsequently the yeast can be collected from the bottom (ale fermentation) or the top (lager fermentation) of the fermenter in order to be used for the next fermentation. While it appears similar to sedimentation in colloidal dispersions, the mechanisms are different.

Water Treatment Process

Flocculation and sedimentation are widely employed in the purification of drinking water as well as in sewage treatment, storm-water treatment and treatment of industrial wastewater streams. Typical treatment processes consist of grates, coagulation, flocculation, sedimentation, granular filtration and disinfection.

Jar Test

The purpose of this test is to select types of coagulant (alum) and also to estimate the optimal dose needed in removing the charged particles that occurred in raw water. Jar test is an experiment to understand the processes of coagulation, flocculation and sedimentation (AWWA, 2011).

Jar test apparatus consists of six batch beakers, and equipped with a paddle mixer for each beaker. In a standard practice, jar test involving of rapid mixing, then follow by slow mixing and later with sedimentation process. An example of the flocculation is shown here.

Deflocculation

Deflocculation is the exact opposite of flocculation. Usually in higher pH ranges in addition to low ionic strengths of soil solutions and domination of alkali metal cations the soil colloidal particles can be dispersed. The additive that prevents the colloids from forming flocs is called a deflocculant. According to the Encyclopedic Dictionary of Polymers deflocculation is "a state or condition of a dispersion of a solid in a liquid in which each solid particle remains independent and unassociated with adjacent particles. A deflocculated suspension shows zero or very low yield value".

Deflocculation can be a problem in wastewater treatment plants as it commonly causes sludge settling problems and deterioration of the effluent quality.

Trickling Filter

A trickling filter is a type of wastewater treatment system first used by Dibden and Clowes It consists of a fixed bed of rocks, lava, coke, gravel, slag, polyurethane foam, sphagnum peat moss, ceramic, or plastic media over which sewage or other wastewater flows downward and causes a layer of microbial slime (biofilm) to grow, covering the bed of media. Aerobic conditions are maintained by splashing, diffusion, and either by forced-air flowing through the bed or natural convection of air if the filter medium is porous.

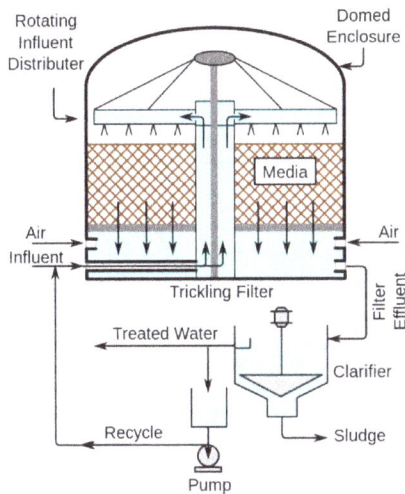

A typical complete trickling filter system

The terms trickle filter, trickling biofilter, biofilter, biological filter and biological trickling filter are often used to refer to a trickling filter. These systems have also been described as roughing filters, intermittent filters, packed media bed filters, alternative septic systems, percolating filters, attached growth processes, and fixed film processes.

Construction

A typical trickling filter is circular and between 10 metres and 20 metres across and between 2 metres to 3 metres deep. A circular wall, often of brick, contains a bed of filter media which in turn rests on a base of under-drains. These under-drains function both to remove liquid passing through the filter media but also to allow the free passage of air up through the filter media. Mounted in the center over the top of the filter media is a spindle supporting two or more horizontal perforated pipes which extend to the edge of the media. The perforations on the pipes are designed to allow an even flow of liquid over the whole area of the media and are also angled so that when liquid flows from the pipes the whole assembly rotates around the central spindle. Settled sewage is delivered to a reservoir at the centre of the spindle via some form of dosing mechanism, often a tipping bucket device on small filters.

Larger filters may be rectangular and the distribution arms may be driven by hydraulic or electrical systems.

Trickling may have a variety of types of filter media used to support the bioi-film. Types of media most commonly used include coke, pumice, plastic matrix material, open-cell polyurethane foam, clinker, gravel, sand and geotextiles. Ideal filter medium optimizes surface area for microbial attachment, wastewater retention time, allows air flow, resists plugging is mechanically robust in all weathers allowing walking access across the filter and does not degrade. Some residential systems require forced aeration units which will increase maintenance and operational costs.

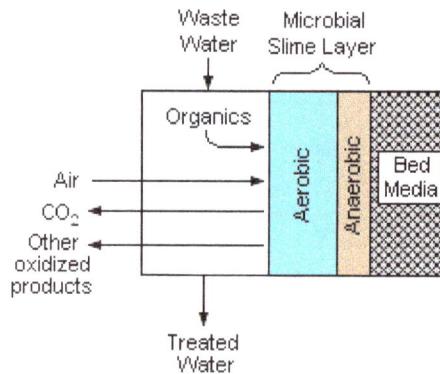

Image 1. A schematic cross-section of the contact face of the bed of media in a trickling filter

Operation

Typically, sewage flow enters at a high level and flows through the primary settlement tank. The supernatant from the tank flows into a dosing device, often a tipping bucket which delivers flow to the arms of the filter. The flush of water flows through the arms and exits through a series of holes pointing at an angle downwards. This propels the arms around distributing the liquid evenly over the surface of the filter media. Most are uncovered (unlike the accompanying diagram) and are freely ventilated to the atmosphere.

Systems can be configured for single-pass use where the treated water is applied to the trickling filter once before being disposed of, or for multi-pass use where a portion of the treated water is cycled back and re-treated via a closed loop. Multi-pass systems result in higher treatment quality and assist in removing Total Nitrogen (TN) levels by promoting nitrification in the aerobic media bed and denitrification in the anaerobic septic tank. Some systems use the filters in two banks operated in series so that the wastewater has two passes through a filter with a sedimentation stage between the two passes. Every few days the filters are switched round to balance the load. This method of treatment can improve nitrification and de-nitrification since much of the carbonaceous oxidative material is removed on the first pass through the filters.

The removal of pollutants from the waste water stream involves both absorption and adsorption of organic compounds and some inorganic species such as nitrite and nitrate ions by the layer of microbial bio film. The filter media is typically chosen to provide a very high surface area to volume. Typical materials are often porous and have considerable internal surface area in addition to the external surface of the medium. Passage of the waste water over the media provides dissolved oxygen which the bio-film layer requires for the biochemical oxidation of the organic compounds and releases carbon dioxide gas, water and other oxidized end products. As the bio film layer thickens, it eventually sloughs off into the liquid flow and subsequently forms part of the secondary sludge. Typically, a trickling filter is followed by a clarifier or sedimentation tank for the separation and removal of the sloughed film. Other filters utilizing higher-density media such as sand, foam and peat moss do not produce a sludge that must be removed, but require forced air blowers and backwashing or an enclosed anaerobic environment.

Biological Processes

The bio-film that develops in a trickling filter may become several millimetres thick and is typically a gelatinous matrix that contains many species of bacteria, cilliates and amoeboid protozoa, annelids, round worms and insect larvae and many other micro fauna. This is very different from many other bio-films which may be less than 1 mm thick. Within the thickness of the biofilm both aerobic and anaerobic zones can exist supporting both oxidative and reductive biological processes. At certain times of year, especially in the spring, rapid growth of organisms in the film may cause the film to be too thick and it may slough off in patches leading to the "spring slough".

Types

Single trickling filters may be used for the treatment of small residential septic tank discharges and very small rural sewage treatment systems. Larger centralized sewage treatment plants typically use many trickling filters in parallel. The treatment of industrial wastewater may involve specialised tricking filters which use plastic media and high flow rates.

Industrial Wastewater Treatment

Wastewaters from a variety of industrial processes have been treated in trickling filters. Such industrial wastewater trickling filters consist of two types:

- Large tanks or concrete enclosures filled with plastic packing or other media.

- Vertical towers filled with plastic packing or other media.

The availability of inexpensive plastic tower packings has led to their use as trickling filter beds in tall towers, some as high as 20 meters. As early as the 1960s, such towers were in use at: the Great Northern Oil's Pine Bend Refinery in Minnesota; the Cities Service Oil Company Trafalgar Refinery in Oakville, Ontario and at a kraft paper mill.

The treated water effluent from industrial wastewater trickling filters is typically processed in a clarifier to remove the sludge that sloughs off the microbial slime layer attached to the trickling filter media as for other trickling filter applications.

Some of the latest trickle filter technology involves aerated biofilters of plastic media in vessels using blowers to inject air at the bottom of the vessels, with either downflow or upflow of the wastewater.

Activated Sludge

The activated sludge process is a process for treating sewage and industrial wastewaters using air and a biological floc composed of bacteria and protozoa.

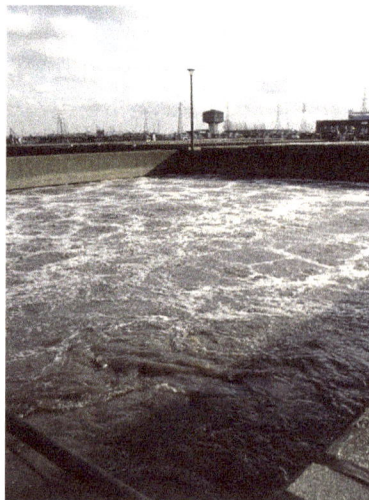

Activated sludge tank at Beckton sewage treatment plant, UK - the white bubbles are due to the diffused air aeration system

Purpose

In a sewage (or industrial wastewater) treatment plant, the activated sludge process is a biological process that can be used for one or several of the following purposes: oxidizing carbonaceous biological matter, oxidizing nitrogenous matter: mainly ammonium and nitrogen in biological matter, removing nutrients (nitrogen and phosphorus).

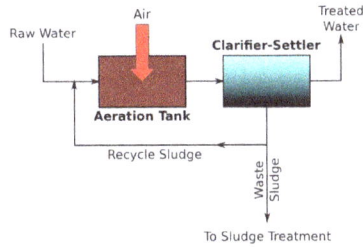

A generalized, schematic diagram of an activated sludge process.

Activated sludge addition (seeding) to a pilot scale membrane bioreactor in Germany

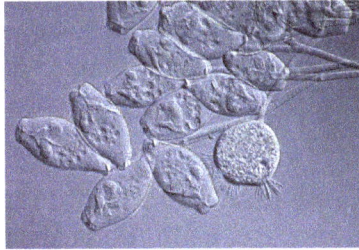

Activated sludge under the microscope

Process Description

The general arrangement of an activated sludge process for removing carbonaceous pollution includes the following items:

- Aeration tank where air (or oxygen) is injected in the mixed liquor.

- Settling tank (usually referred to as "final clarifier" or "secondary settling tank") to allow the biological flocs (the sludge blanket) to settle, thus separating the biological sludge from the clear treated water.

Treatment of nitrogenous matter or phosphate involves additional steps where the mixed liquor is left in anoxic condition (meaning that there is no residual dissolved oxygen).

Bioreactor and Final Clarifier

The process involves air or oxygen being introduced into a mixture of screened, and primary treated sewage or industrial wastewater (wastewater) combined with organisms to develop a biological floc which reduces the organic content of the sewage. This material, which in healthy sludge is a brown floc, is largely composed of saprotrophic bacteria but also has an important protozoan flora component mainly composed of amoebae, Spirotrichs, Peritrichs including Vorticellids and a range of other filter feeding species. Other important constituents include motile and sedentary Rotifers. In poorly managed activated sludge, a range of mucilaginous filamentous bacteria can develop including *Sphaerotilus natans* which produces a sludge that is difficult to settle and can result in the sludge blanket decanting over the weirs in the settlement tank to severely contaminate the final effluent quality. This material is often described as sewage fungus but true fungal communities are relatively uncommon.

The combination of wastewater and biological mass is commonly known as *mixed liquor*. In all activated sludge plants, once the wastewater has received sufficient treatment, excess mixed liquor is discharged into settling tanks and the treated supernatant is run off to undergo further treatment before discharge. Part of the settled material, the sludge, is returned to the head of the aeration system to re-seed the new wastewater entering the tank. This fraction of the floc is called *return activated sludge* (R.A.S.).

The space required for a sewage treatment plant can be reduced by using a membrane bioreactor to remove some wastewater from the mixed liquor prior to treatment. This results in a more concentrated waste product that can then be treated using the activated sludge process.

Many sewage treatment plants use axial flow pumps to transfer nitrified mixed liquor from the aeration zone to the anoxic zone for denitrification. These pumps are often referred to as internal mixed liquor recycle pumps (IMLR pumps). The raw sewage, the RAS, and the nitrified mixed liquor are mixed by submersible mixers in the anoxic zones in order to achieve denitrification.

Sludge Production

Activated sludge is also the name given to the active biological material produced by activated sludge plants. Excess sludge is called "surplus activated sludge" or "waste activated sludge" and is removed from the treatment process to keep the ratio of biomass to food supplied in the wastewater in balance. This sewage sludge is usually mixed with primary sludge from the primary clarifiers and undergoes further sludge treatment for example by anaerobic digestion, followed by thickening, dewatering, composting and land application.

The amount of sewage sludge produced from the activated sludge process is directly proportional to the amount of wastewater treated. The total sludge production consists of the sum of primary sludge from the primary sedimentation tanks as well as waste activated sludge from the bioreactors. The activated sludge process produces about 70–100 kg/ML of waste activated sludge (that is kg of dry solids produced per ML of wastewater treated; one mega litre (ML) is 10^3 m^3). A value of 80 kg/ML is regarded as being typical. In addition, about 110–170 kg/ML of primary sludge is produced in the primary sedimentation tanks which most - but not all - of the activated sludge process configurations use.

A variant of the activated sludge process is the Nereda process where aerobic granular sludge is developed by applying specific process conditions that favour slow growing organisms.

Process Control

The general method to do this is to monitor sludge blanket level, SVI (Sludge Volume Index), MCRT (Mean Cell Residence Time), F/M (Food to Microorganism), as well as the biota of the activated sludge and the major nutrients DO (Dissolved oxygen), nitrogen, phosphate, BOD (Biochemical oxygen demand), and COD (Chemical oxygen demand).

In the reactor/aerator + clarifier system:

- The sludge blanket is measured from the bottom of the clarifier to the level of settled solids in the clarifier's water column; this, in large plants, can be done up to three times a day.

- The SVI is the volume of settled sludge in milliliters occupied by 1 gram of dry sludge solids after 30 minutes of settling in a 1000 milliliter graduated cylinder.

- The MCRT is the total mass (lbs) of mixed liquor suspended solids in the aerator and clarifier divided by the mass flow rate (lbs/day) of mixed liquor suspended solids leaving as WAS and final effluent.

- The F/M is the ratio of food fed to the microorganisms each day to the mass of microorganisms held under aeration. Specifically, it is the amount of BOD fed to the aerator (lbs/day) divided by the amount (lbs) of Mixed Liquor Volatile Suspended Solids (MLVSS) under aeration. Note: Some references use MLSS (Mixed Liquor Suspended Solids) for expedience, but MLVSS is considered more accurate for the measure of microorganisms. Again, due to expedience, COD is generally used, in lieu of BOD, as BOD takes five days for results.

Based on these control methods, the amount of settled solids in the mixed liquor can be varied by wasting activated sludge (WAS) or returning activated sludge (RAS).

Types of Plants

There are a variety of types of activated sludge plants. These include:

Package Plants

There are a wide range of types of package plants, often serving small communities or industrial plants that may use hybrid treatment processes often involving the use of aerobic sludge to treat the incoming sewage. In such plants the primary settlement stage of treatment may be omitted. In these plants, a biotic floc is created which provides the required substrate. Package plants are designed and fabricated by specialty engineering firms in dimensions that allow for their transportation to the job site in public highways, typically width and height of 12 by 12 feet. Length varies with capacity with larger plants being fabricated in pieces and welded on site. Steel is preferred over synthetic materials (e.g., plastic) for its durability.

Package plants are commonly variants of extended aeration, to promote the 'fit & forget' approach required for small communities without dedicated operational staff. There are various standards to assist with their design.

Oxidation Ditch

In some areas, where more land is available, sewage is treated in large round or oval ditches with one or more horizontal aerators typically called brush or disc aerators which drive the mixed liquor around the ditch and provide aeration. These are oxidation ditches, often referred to by manufacturer's trade names such as Pasveer, Orbal, or Carrousel. They have the advantage that they are relatively easy to maintain and are resilient to shock loads that often occur in smaller communities (i.e. at breakfast time and in the evening).

Oxidation ditches are installed commonly as 'fit & forget' technology, with typical design parameters of a hydraulic retention time of 24 – 48 hours, and a sludge age of 12 – 20 days. This compares with nitrifying activated sludge plants having a retention time of 8 hours, and a sludge age of 8 – 12 days.

Deep Shaft/Vertical Treatment

Where land is in short supply sewage may be treated by injection of oxygen into a pressured return sludge stream which is injected into the base of a deep columnar tank buried in the ground. Such shafts may be up to 100 metres deep and are filled with sewage liquor. As the sewage rises the oxygen forced into solution by the pressure at the base of the shaft breaks out as molecular oxygen providing a highly efficient source of oxygen for the activated sludge biota. The rising oxygen and injected return sludge provide the physical mechanism for mixing of the sewage and sludge. Mixed sludge and sewage is decanted at the surface and separated into supernatant and sludge components. The efficiency of deep shaft treatment can be high.

Surface aerators are commonly quoted as having an aeration efficiency of 0.5 - 1.5 kg O_2/kWh, diffused aeration as 1.5 - 2.5 kg O_2/KWh. Deep Shaft claims 5 – 8 kg O_2/kWh.

However, the costs of construction are high. Deep Shaft has seen the greatest uptake in Japan, because of the land area issues. Deep Shaft was developed by ICI, as a spin-off from their Pruteen process. In the UK it is found at three sites: Tilbury, Anglian water, treating a wastewater with a high industrial contribution; Southport, United Utilities, because of land space issues; and Billingham, ICI, again treating industrial effluent, and built (after the Tilbury shafts) by ICI to help the agent sell more.

DeepShaft is a patented, licensed, process. The licensee has changed several times and currently (2015) Noram Engineering sells it.

Surface-aerated Basins

Most biological oxidation processes for treating industrial wastewaters have in common the use of oxygen (or air) and microbial action. Surface-aerated basins achieve 80 to 90% removal of BOD with retention times of 1 to 10 days. The basins may range in depth from 1.5 to 5.0 metres and utilize motor-driven aerators floating on the surface of the wastewater.

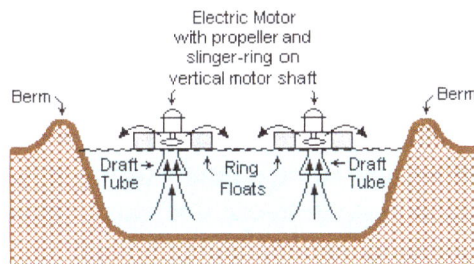

A TYPICAL SURFACE – AERATED BASIN
Note: The ring floats are tethered to posts on the berms.

A Typical Surface-Aerated Basing (using motor-driven floating aerators)

In an aerated basin system, the aerators provide two functions: they transfer air into the basins required by the biological oxidation reactions, and they provide the mixing required for dispersing the air and for contacting the reactants (that is, oxygen, wastewater and microbes). Typically, the floating surface aerators are rated to deliver the amount of air equivalent to 1.8 to 2.7 kg O_2/kWh. However, they do not provide as good mixing as is normally achieved in activated sludge systems and therefore aerated basins do not achieve the same performance level as activated sludge units.

Biological oxidation processes are sensitive to temperature and, between 0 °C and 40 °C, the rate of biological reactions increase with temperature. Most surface aerated vessels operate at between 4 °C and 32 °C.

Sequencing Batch Reactors (SBRs)

Sequencing batch reactors (SBRs) treat wastewater in batches within the same vessel. This means that the bioreactor and final clarifier are not separates in space but in a timed sequence. The installation consists of at least two identically equipped tanks with a common inlet, which can be switched between them. While one tank is in settle/decant mode the other is aerating and filling.

Aeration Methods

Diffused Aeration

Sewage liquor is run into deep tanks with diffuser grid aeration systems that are attached to the floor. These are like the diffused airstone used in tropical fish tanks but on a much larger scale. Air is pumped through the blocks and the curtain of bubbles formed both oxygenates the liquor and also provides the necessary mixing action. Where capacity is limited or the sewage is unusually strong or difficult to treat, oxygen may be used instead of air. Typically, the air is generated by some type of blower or compressor.

Surface Aerators (Cones)

Vertically mounted tubes of up to 1 metre diameter extending from just above the base of a deep concrete tank to just below the surface of the sewage liquor. A typical shaft might be 10 metres high. At the surface end the tube is formed into a cone with helical vanes attached to the inner surface. When the tube is rotated, the vanes spin liquor up and out of the cones drawing new sewage liquor from the base of the tank. In many works each cone is located in a separate cell that can be isolated from the remaining cells if required for maintenance. Some works may have two cones to a cell and some large works may have 4 cones per cell.

Pure Oxygen Aeration

Pure oxygen activated sludge aeration systems are sealed-tank reactor vessels with surface aerator type impellers mounted within the tanks at the oxygen atmosphere-mixed liquor surface interface. The amount of oxygen entrainment, or DO (Dissolved Oxygen), can be controlled by a weir adjusted level control, and a vent gas oxygen controlled oxygen feed valve. Oxygen is generated on site by cryogenic distillation of air, pressure swing adsorption, or other methods. These systems are used where wastewater plant space is at a premium and high sewage throughput is required as high energy costs are involved in purifying oxygen.

Recent Developments

A new development of the activated sludge process is the Nereda process which produces a granular sludge that settles very well (the sludge volume index is reduced from

200-300 to 40 mL/g). A new process reactor system is created to take advantage of this quick settling sludge and is integrated into the aeration tank instead of having a separate unit outside. About 30 Nereda wastewater treatment plants worldwide are operational, under construction or under design, varying in size from 5,000 up to 858,000 person equivalent.

History

The activated sludge process was discovered in 1913 in the United Kingdom by two engineers, Edward Ardern and W.T. Lockett, who were conducting research for the Manchester Corporation Rivers Department at Davyhulme Sewage Works. This development led to arguably the single most significant improvement in public health and the environment during the course of the century.

The Davyhulme Sewage Works Laboratory, where the activated sludge process was developed in the early 20th century.

In 1912, Dr. Gilbert Fowler, a scientist at the University of Manchester, observed experiments being conducted at the Lawrence Experiment Station at Massachusetts involving the aeration of sewage in a bottle that had been coated with algae. Fowler's engineering colleagues, Ardern and Lockett, experimented on treating sewage in a draw-and-fill reactor, which produced a highly treated effluent. They aerated the waste-water continuously for about a month and were able to achieve a complete nitrification of the sample material. Believing that the sludge had been activated (in a similar manner to activated carbon) the process was named *activated sludge*. Not until much later was it realized that what had actually occurred was a means to concentrate biological organisms, decoupling the liquid retention time (ideally, low, for a compact treatment system) from the solids retention time (ideally, fairly high, for an effluent low in BOD_5 and ammonia.)

Their results were published in their seminal 1914 paper, and the first full-scale continuous-flow system was installed at Worcester two years later. In the aftermath of the First World War the new treatment method spread rapidly, especially to the USA, Denmark, Germany and Canada. By the late 1930s, the activated sludge treatment became a well-known biological wastewater treatment process in those countries where sewer systems and sewage treatment plants were common.

References

- Tchobanoglous, G., Burton, F.L., and Stensel, H.D. (2003). Wastewater Engineering (Treatment Disposal Reuse) / Metcalf & Eddy, Inc. (4th ed.). McGraw-Hill Book Company. ISBN 0-07-041878-0.

- Crittenden, John; Trussell, Rhodes; Hand, David; Howe, Kerry and Tchobanoglous, George (2005). Water Treatment Principles and Design, Edition 2. John Wiley and Sons. New Jersey. ISBN 0-471-11018-3

- Knorr, Erik Voigt, Henry Jaeger, Dietrich (2012). Securing Safe Water Supplies : comparison of applicable technologies (Online-Ausg. ed.). Oxford: Academic Press. p. 33. ISBN 0124058868. Retrieved 4 March 2015.

- Grabowski, Andrej (2010). Electromembrane desalination processes for production of low conductivity water. Berlin: Logos-Verl. ISBN 3832527141. Retrieved 4 March 2015.

- Weber, Walter J. (1972). Physicochemical Processes for Water Quality Control. New York: John Wiley & Sons. p. 320. ISBN 0-471-92435-0.

- Hubbard, Arthur T. (2004). Encyclopedia of Surface and Colloid Science. CRC Press. p. 4230. ISBN 0-8247-0759-1. Retrieved 2007-11-13.

- Fox, Patrick F. (1999). Cheese Volume 1: Chemistry, Physics, and Microbiology (2nd ed.). Gaithersburg, Maryland: Aspen Publishers. pp. 144–150. ISBN 978-0-8342-1378-4.

- Fox, Patrick F. (2004). Cheese - Chemistry, Physics and Microbiology (3rd Edition). Elsevier. p. 72. ISBN 978-0-12-263653-0.

- Gooch, Dr Jan W., ed. (2007-01-01). Deflocculation. Springer New York. pp. 265–265. doi:10.1007/978-0-387-30160-0_3313. ISBN 978-0-387-31021-3.

- Wastewater engineering : treatment and reuse (4th ed.). Metcalf & Eddy, Inc., McGraw Hill, USA. 2003. p. 1456. ISBN 0-07-112250-8.

Industrial Waste Treatment: Various Techniques

Tools and techniques are an important component of any field of study. Techniques such as incineration, pyrolysis and landfill are discussed in detail in this chapter. Incineration is a process that involves the burning of organic substances contained in waste materials whereas recycling is the method of converting waste into reusable objects.

Incineration

Incineration is a waste treatment process that involves the combustion of organic substances contained in waste materials. Incineration and other high-temperature waste treatment systems are described as "thermal treatment". Incineration of waste materials converts the waste into ash, flue gas, and heat. The ash is mostly formed by the inorganic constituents of the waste, and may take the form of solid lumps or particulates carried by the flue gas. The flue gases must be cleaned of gaseous and particulate pollutants before they are dispersed into the atmosphere. In some cases, the heat generated by incineration can be used to generate electric power.

The Spittelau incineration plant in Vienna, Austria, designed by Friedensreich Hundertwasser.

Incineration with energy recovery is one of several waste-to-energy (WtE) technologies such as gasification, pyrolysis and anaerobic digestion. While incineration and gasification technologies are similar in principle, the energy product from incineration is high-temperature heat whereas combustible gas is often the main energy product from gasification. Incineration and gasification may also be implemented without energy and materials recovery.

SYSAV incineration plant in Malmö, Sweden, capable of handling 25 metric tons (28 short tons) per hour of household waste. To the left of the main stack, a new identical oven line is under construction (March 2007).

In several countries, there are still concerns from experts and local communities about the environmental effect of incinerators.

In some countries, incinerators built just a few decades ago often did not include a materials separation to remove hazardous, bulky or recyclable materials before combustion. These facilities tended to risk the health of the plant workers and the local environment due to inadequate levels of gas cleaning and combustion process control. Most of these facilities did not generate electricity.

Incinerators reduce the solid mass of the original waste by 80–85% and the volume (already compressed somewhat in garbage trucks) by 95–96%, depending on composition and degree of recovery of materials such as metals from the ash for recycling. This means that while incineration does not completely replace landfilling, it significantly reduces the necessary volume for disposal. Garbage trucks often reduce the volume of waste in a built-in compressor before delivery to the incinerator. Alternatively, at landfills, the volume of the uncompressed garbage can be reduced by approximately 70% by using a stationary steel compressor, albeit with a significant energy cost. In many countries, simpler waste compaction is a common practice for compaction at landfills.

Incineration has particularly strong benefits for the treatment of certain waste types in niche areas such as clinical wastes and certain hazardous wastes where pathogens and toxins can be destroyed by high temperatures. Examples include chemical multi-product plants with diverse toxic or very toxic wastewater streams, which cannot be routed to a conventional wastewater treatment plant.

Waste combustion is particularly popular in countries such as Japan where land is a scarce resource. Denmark and Sweden have been leaders in using the energy generated from incineration for more than a century, in localised combined heat and power facilities supporting district heating schemes. In 2005, waste incineration produced 4.8% of the electricity consumption and 13.7% of the total domestic heat consumption in Denmark. A number of other European countries rely heavily on incineration for handling municipal waste, in particular Luxembourg, the Netherlands, Germany, and France.

History

The first UK incinerators for waste disposal were built in Nottingham by Manlove, Alliott & Co. Ltd. in 1874 to a design patented by Albert Fryer. They were originally known as destructors.

Manlove, Alliott & Co. Ltd. 1894 destructor furnace at Cambridge Museum of Technology

The first US incinerator was built in 1885 on Governors Island in New York, NY. The first facility in the Czech Republic was built in 1905 in Brno.

The Encyclopaedia Britannica Eleventh Edition contains a detailed contemporaneous description of the history and design of destructors. For more information, see that entry.

Technology

An incinerator is a furnace for burning waste. Modern incinerators include pollution mitigation equipment such as flue gas cleaning. There are various types of incinerator plant design: moving grate, fixed grate, rotary-kiln, and fluidised bed.

Burn Pile

The burn pile is one of the simplest and earliest forms of waste disposal, essentially consisting of a mound of combustible materials piled on open ground and set on fire.

Burn piles can and have spread uncontrolled fires, for example if wind blows burning material off the pile into surrounding combustible grasses or onto buildings. As interior structures of the pile are consumed, the pile can shift and collapse, spreading the burn area. Even in a situation of no wind, small lightweight ignited embers can lift off the pile via convection, and waft through the air into grasses or onto buildings, igniting them.

A typical small burn pile in a garden.

Burn Barrel

The burn barrel is a somewhat more controlled form of private waste incineration, containing the burning material inside a metal barrel, with a metal grating over the exhaust. The barrel prevents the spread of burning material in windy conditions, and as the combustibles are reduced they can only settle down into the barrel. The exhaust grating helps to prevent the spread of burning embers. Typically steel 55-US-gallon (210 L) drums are used as burn barrels, with air vent holes cut or drilled around the base for air intake. Over time, the very high heat of incineration causes the metal to oxidize and rust, and eventually the barrel itself is consumed by the heat and must be replaced.

Private burning of dry cellulosic/paper products is generally clean-burning, producing no visible smoke, but plastics in household waste can cause private burning to create a public nuisance, generating acrid odors and fumes that make eyes burn and water. Most urban communities ban burn barrels, and certain rural communities may have prohibitions on open burning, especially those home to many residents not familiar with this common rural practice.

As of 2006 in the United States, private rural household or farm waste incineration of small quantities was typically permitted so long as it is not a nuisance to others, does not pose a risk of fire such as in dry conditions, and the fire does not produce dense, noxious smoke. A handful of states, such as New York, Minnesota, and Wisconsin, have laws or regulations either banning or strictly regulating open burning due to health and nuisance effects. People intending to burn waste may be required to contact a state agency in advance to check current fire risk and conditions, and to alert officials of the controlled fire that will occur.

Moving Grate

The typical incineration plant for municipal solid waste is a moving grate incinerator. The moving grate enables the movement of waste through the combustion chamber to be optimised to allow a more efficient and complete combustion. A single moving grate boiler can handle up to 35 metric tons (39 short tons) of waste per hour, and can operate 8,000 hours per year with only one scheduled stop for inspection and maintenance of about one month's duration. Moving grate incinerators are sometimes referred to as Municipal Solid Waste Incinerators (MSWIs).

Control room of a typical moving grate incinerator overseeing two boiler lines

The waste is introduced by a waste crane through the "throat" at one end of the grate, from where it moves down over the descending grate to the ash pit in the other end. Here the ash is removed through a water lock.

Municipal solid waste in the furnace of a moving grate incinerator capable of handling 15 metric tons (17 short tons) of waste per hour. The holes in the grate elements supplying the primary combustion air are visible.

Part of the combustion air (primary combustion air) is supplied through the grate from below. This air flow also has the purpose of cooling the grate itself. Cooling is important for the mechanical strength of the grate, and many moving grates are also water-cooled internally.

Secondary combustion air is supplied into the boiler at high speed through nozzles over the grate. It facilitates complete combustion of the flue gases by introducing turbulence for better mixing and by ensuring a surplus of oxygen. In multiple/stepped hearth incinerators, the secondary combustion air is introduced in a separate chamber downstream the primary combustion chamber.

According to the European Waste Incineration Directive, incineration plants must be designed to ensure that the flue gases reach a temperature of at least 850 °C (1,560 °F) for 2 seconds in order to ensure proper breakdown of toxic organic substances. In order to comply with this at all times, it is required to install backup auxiliary burners (often fueled by oil), which are fired into the boiler in case the heating value of the waste becomes too low to reach this temperature alone.

The flue gases are then cooled in the superheaters, where the heat is transferred to steam, heating the steam to typically 400 °C (752 °F) at a pressure of 40 bars (580 psi) for the electricity generation in the turbine. At this point, the flue gas has a temperature of around 200 °C (392 °F), and is passed to the flue gas cleaning system.

In Scandinavia, scheduled maintenance is always performed during summer, where the demand for district heating is low. Often, incineration plants consist of several separate 'boiler lines' (boilers and flue gas treatment plants), so that waste can continue to be received at one boiler line while the others are undergoing maintenance, repair, or upgrading.

Fixed Grate

The older and simpler kind of incinerator was a brick-lined cell with a fixed metal grate over a lower ash pit, with one opening in the top or side for loading and another opening in the side for removing incombustible solids called clinkers. Many small incinerators formerly found in apartment houses have now been replaced by waste compactors.

Rotary-kiln

The rotary-kiln incinerator is used by municipalities and by large industrial plants. This design of incinerator has 2 chambers: a primary chamber and secondary chamber. The primary chamber in a rotary kiln incinerator consist of an inclined refractory lined cylindrical tube. The inner refractory lining serves as sacrificial layer to protect the kiln structure. This refractory layer needs to be replaced from time to time. Movement of the cylinder on its axis facilitates movement of waste. In the primary chamber, there is conversion of solid fraction to gases, through volatilization, destructive distillation and partial combustion reactions. The secondary chamber is necessary to complete gas phase combustion reactions.

The clinkers spill out at the end of the cylinder. A tall flue-gas stack, fan, or steam jet supplies the needed draft. Ash drops through the grate, but many particles are carried along with the hot gases. The particles and any combustible gases may be combusted in an "afterburner".

Fluidized Bed

A strong airflow is forced through a sandbed. The air seeps through the sand until a point is reached where the sand particles separate to let the air through and mixing and churning occurs, thus a fluidized bed is created and fuel and waste can now be introduced. The sand with the pre-treated waste and/or fuel is kept suspended on pumped air currents and takes on a fluid-like character. The bed is thereby violently mixed and agitated keeping small inert particles and air in a fluid-like state. This allows all of the mass of waste, fuel and sand to be fully circulated through the furnace.

Specialized Incineration

Furniture factory sawdust incinerators need much attention as these have to handle resin powder and many flammable substances. Controlled combustion, burn back prevention systems are essential as dust when suspended resembles the fire catch phenomenon of any liquid petroleum gas.

Use of Heat

The heat produced by an incinerator can be used to generate steam which may then be used to drive a turbine in order to produce electricity. The typical amount of net energy that can be produced per tonne municipal waste is about 2/3 MWh of electricity and 2 MWh of district heating. Thus, incinerating about 600 metric tons (660 short tons) per day of waste will produce about 400 MWh of electrical energy per day (17 MW of electrical power continuously for 24 hours) and 1200 MWh of district heating energy each day.

Pollution

Incineration has a number of outputs such as the ash and the emission to the atmosphere of flue gas. Before the flue gas cleaning system, if installed, the flue gases may contain particulate matter, heavy metals, dioxins, furans, sulfur dioxide, and hydrochloric acid. If plants have inadequate flue gas cleaning, these outputs may add a significant pollution component to stack emissions.

In a study from 1997, Delaware Solid Waste Authority found that, for same amount of produced energy, incineration plants emitted fewer particles, hydrocarbons and less SO_2, HCl, CO and NO_x than coal-fired power plants, but more than natural gas–fired power plants. According to Germany's Ministry of the Environment, waste incinerators reduce the amount of some atmospheric pollutants by substituting power produced by coal-fired plants with power from waste-fired plants.

Gaseous Emissions

Dioxin and Furans

The most publicized concerns from environmentalists about the incineration of municipal solid wastes (MSW) involve the fear that it produces significant amounts of dioxin and furan emissions. Dioxins and furans are considered by many to be serious health hazards. The EPA announced in 2012 that the safe limit for human oral consumption is 0.7 picograms Toxic Equivalence (TEQ) per kilogram bodyweight per day, which works out to 17 billionths of a gram for a 150 lb person per year.

In 2005, The Ministry of the Environment of Germany, where there were 66 incinerators at that time, estimated that "...whereas in 1990 one third of all dioxin emissions in Germany came from incineration plants, for the year 2000 the figure was less than 1%. Chimneys and tiled stoves in private households alone discharge approximately 20 times more dioxin into the environment than incineration plants."

According to the United States Environmental Protection Agency, the combustion percentages of the total dioxin and furan inventory from all known and estimated sources in the U.S. (not only incineration) for each type of incineration are as follows: 35.1% backyard barrels; 26.6% medical waste; 6.3% municipal wastewater treatment sludge; 5.9% municipal waste combustion; 2.9% industrial wood combustion. Thus, the controlled combustion of waste accounted for 41.7% of the total dioxin inventory.

In 1987, before the governmental regulations required the use of emission controls, there was a total of 8,905.1 grams (314.12 oz) Toxic Equivalence (TEQ) of dioxin emissions from US municipal waste combustors. Today, the total emissions from the plants are 83.8 grams (2.96 oz) TEQ annually, a reduction of 99%.

Backyard barrel burning of household and garden wastes, still allowed in some rural areas, generates 580 grams (20 oz) of dioxins annually. Studies conducted by the US-EPA demonstrated that the emissions from just one family using a burn barrel produced more emissions than an incineration plant disposing of 200 metric tons (220 short tons) of waste per day by 1997 and five times that by 2007 due to increased chemicals in household trash and decreased emissions by municipal incinerators using better technology.

However, the same researchers found that their original estimates for the burn barrel were high, and that the incineration plant used for comparison represented a theoretical 'clean' plant rather than any existing facility. Their later studies found that burn barrels produced a median of 24.95 nanograms TEQ per lb garbage burned, so that a family burning 5 lbs of trash per day, or 1825 lbs per year, produces a total of 0.0455 mg TEQ per year, and that the equivalent number of burn barrels for the 83.8 grams (2.96 oz) of the 251 municipal waste combustors inventoried by the EPA in the U.S. in 2000, is 1,841,700, or on average, 7337 family burn barrels per municipal waste incinerator.

Most of the improvement in U.S. dioxin emissions has been for large-scale municipal waste incinerators. As of the year 2000, although small-scale incinerators (those with a daily capacity of less than 250 tons) processed only 9% of the total waste combusted, these produced 83% of the dioxins and furans emitted by municipal waste combustion.

Dioxin Cracking Methods and Limitations

The breakdown of dioxin requires exposure of the molecular ring to a sufficiently high temperature so as to trigger thermal breakdown of the strong molecular bonds holding it together. Small pieces of fly ash may be somewhat thick, and too brief an exposure to high temperature may only degrade dioxin on the surface of the ash. For a large volume air chamber, too brief an exposure may also result in only some of the exhaust gases reaching the full breakdown temperature. For this reason there is also a time element to the temperature exposure to ensure heating completely through the thickness of the fly ash and the volume of waste gases.

There are trade-offs between increasing either the temperature or exposure time. Generally where the molecular breakdown temperature is higher, the exposure time for heating can be shorter, but excessively high temperatures can also cause wear and damage to other parts of the incineration equipment. Likewise the breakdown temperature can be lowered to some degree but then the exhaust gases would require a greater lingering period of perhaps several minutes, which would require large/long treatment chambers that take up a great deal of treatment plant space.

A side effect of breaking the strong molecular bonds of dioxin is the potential for breaking the bonds of nitrogen gas (N_2) and oxygen gas (O_2) in the supply air. As the exhaust flow cools, these highly reactive detached atoms spontaneously reform bonds into reactive oxides such as NO_x in the flue gas, which can result in smog formation and acid rain if they were released directly into the local environment. These reactive oxides must be further neutralized with selective catalytic reduction (SCR) or selective non-catalytic reduction.

Dioxin Cracking in Practice

The temperatures needed to break down dioxin are typically not reached when burning plastics outdoors in a burn barrel or garbage pit, causing high dioxin emissions as mentioned above. While plastic does usually burn in an open-air fire, the dioxins remain after combustion and either float off into the atmosphere, or may remain in the ash where it can be leached down into groundwater when rain falls on the ash pile. Fortunately, dioxin and furan compounds bond very strongly to solid surfaces and are not dissolved by water, so leaching processes are limited to the first few milimeters below the ash pile. The gas-phase dioxins can be substantially destroyed using catalysts, some of which can be present as part of the fabric filter bag structure.

Modern municipal incinerator designs include a high-temperature zone, where the flue gas is sustained at a temperature above 850 °C (1,560 °F) for at least 2 seconds before it is cooled down. They are equipped with auxiliary heaters to ensure this at all times. These are often fueled by oil or natural gas, and are normally only active for a very small fraction of the time. Further, most modern incinerators utilize fabric filters (often with

Teflon membranes to enhance collection of sub-micron particles) which can capture dioxins present in or on solid particles.

For very small municipal incinerators, the required temperature for thermal break-down of dioxin may be reached using a high-temperature electrical heating element, plus a selective catalytic reduction stage.

Although dioxins and furans may be destroyed by combustion, their reformation by a process known as 'de novo synthesis' as the emission gases cool is a probable source of the dioxins measured in emission stack tests from plants that have high combustion temperatures held at long residence times.

CO_2

As for other complete combustion processes, nearly all of the carbon content in the waste is emitted as CO_2 to the atmosphere. MSW contains approximately the same mass fraction of carbon as CO_2 itself (27%), so incineration of 1 ton of MSW produces approximately 1 ton of CO_2.

If the waste was landfilled, 1 ton of MSW would produce approximately 62 cubic metres (2,200 cu ft) methane via the anaerobic decomposition of the biodegradable part of the waste. Since the global warming potential of methane is 34 and the weight of 62 cubic meters of methane at 25 degrees Celsius is 40.7 kg, this is equivalent to 1.38 ton of CO_2, which is more than the 1 ton of CO_2 which would have been produced by incineration. In some countries, large amounts of landfill gas are collected. Still the global warming potential of the landfill gas emitted to atmosphere is significant. In the US it was estimated that the global warming potential of the emitted landfill gas in 1999 was approximately 32% higher than the amount of CO_2 that would have been emitted by incineration. Since this study, the global warming potential estimate for methane has been increased from 21 to 35, which alone would increase this estimate to almost the triple GWP effect compared to incineration of the same waste.

In addition, nearly all biodegradable waste has biological origin. This material has been formed by plants using atmospheric CO_2 typically within the last growing season. If these plants are regrown the CO_2 emitted from their combustion will be taken out from the atmosphere once more.

Such considerations are the main reason why several countries administrate incineration of biodegradable waste as renewable energy. The rest – mainly plastics and other oil and gas derived products – is generally treated as non-renewables.

Different results for the CO_2 footprint of incineration can be reached with different assumptions. Local conditions (such as limited local district heating demand, no fossil fuel generated electricity to replace or high levels of aluminium in the waste stream) can decrease the CO_2 benefits of incineration. The methodology and other assumptions may also influence the results significantly. For example, the methane emissions from landfills occurring at a later date may be neglected or given less weight, or biodegradable waste may not be considered CO_2 neutral. A study by Eunomia Research and Consulting in 2008 on potential waste treatment technologies in London demonstrated that by applying several of these (according to the authors) unusual assumptions the average existing incineration plants performed poorly for CO_2 balance compared to the theoretical potential of other emerging waste treatment technologies.

Other Emissions

Other gaseous emissions in the flue gas from incinerator furnaces include nitrogen oxides, sulfur dioxide, hydrochloric acid, heavy metals, and fine particles. Of the heavy metals, mercury is a major concern due to its toxicity and high volatility, as essentially all mercury in the municipal waste stream may exit in emissions if not removed by emission controls.

The steam content in the flue may produce visible fume from the stack, which can be perceived as a visual pollution. It may be avoided by decreasing the steam content by flue-gas condensation and reheating, or by increasing the flue gas exit temperature well above its dew point. Flue-gas condensation allows the latent heat of vaporization of the water to be recovered, subsequently increasing the thermal efficiency of the plant.

Flue-gas Cleaning

The quantity of pollutants in the flue gas from incineration plants may or may not be reduced by several processes, depending on the plant.

Electrodes inside electrostatic precipitator

Particulate is collected by particle filtration, most often electrostatic precipitators (ESP) and/or baghouse filters. The latter are generally very efficient for collecting fine particles. In an investigation by the Ministry of the Environment of Denmark in 2006, the average particulate emissions per energy content of incinerated waste from 16 Danish incinerators were below 2.02 g/GJ (grams per energy content of the incinerated waste). Detailed measurements of fine particles with sizes below 2.5 micrometres ($PM_{2.5}$) were performed on three of the incinerators: One incinerator equipped with an ESP for particle filtration emitted 5.3 g/GJ fine particles, while two incinerators equipped with baghouse filters emitted 0.002 and 0.013 g/GJ $PM_{2.5}$. For ultra fine particles ($PM_{1.0}$), the numbers were 4.889 g/GJ $PM_{1.0}$ from the ESP plant, while emissions of 0.000 and 0.008 g/GJ $PM_{1.0}$ were measured from the plants equipped with baghouse filters.

Acid gas scrubbers are used to remove hydrochloric acid, nitric acid, hydrofluoric acid, mercury, lead and other heavy metals. The efficiency of removal will depend on the specific equipment, the chemical composition of the waste, the design of the plant, the chemistry of reagents, and the ability of engineers to optimize these conditions, which may conflict for different pollutants. For example, mercury removal by wet scrubbers is considered coincidental and may be less than 50%. Basic scrubbers remove sulfur dioxide, forming gypsum by reaction with lime.

Waste water from scrubbers must subsequently pass through a waste water treatment plant.

Sulfur dioxide may also be removed by dry desulfurisation by injection limestone slurry into the flue gas before the particle filtration.

NO_x is either reduced by catalytic reduction with ammonia in a catalytic converter (selective catalytic reduction, SCR) or by a high-temperature reaction with ammonia in the furnace (selective non-catalytic reduction, SNCR). Urea may be substituted for ammonia as the reducing reagent but must be supplied earlier in the process so that it can hydrolyze into ammonia. Substitution of urea can reduce costs and potential hazards associated with storage of anhydrous ammonia.

Heavy metals are often adsorbed on injected active carbon powder, which is collected by particle filtration.

Solid Outputs

Operation of an incinerator aboard an aircraft carrier

Incineration produces fly ash and bottom ash just as is the case when coal is combusted. The total amount of ash produced by municipal solid waste incineration ranges from 4 to 10% by volume and 15–20% by weight of the original quantity of waste, and the fly ash amounts to about 10–20% of the total ash. The fly ash, by far, constitutes more of a potential health hazard than does the bottom ash because the fly ash often contain high concentrations of heavy metals such as lead, cadmium, copper and zinc as well as small amounts of dioxins and furans. The bottom ash seldom contain significant levels of heavy metals. In testing over the past decade, no ash from an incineration plant in the USA has ever been determined to be a hazardous waste. At present although some historic samples tested by the incinerator operators' group would meet the being eco-toxic criteria at present the EA say "we have agreed" to regard incinerator bottom ash as "non-hazardous" until the testing programme is complete.

Other Pollution Issues

Odor pollution can be a problem with old-style incinerators, but odors and dust are extremely well controlled in newer incineration plants. They receive and store the waste in an enclosed area with a negative pressure with the airflow being routed through the boiler which prevents unpleasant odors from escaping into the atmosphere. However, not all plants are implemented this way, resulting in inconveniences in the locality.

An issue that affects community relationships is the increased road traffic of waste collection vehicles to transport municipal waste to the incinerator. Due to this reason, most incinerators are located in industrial areas. This problem can be avoided to an extent through the transport of waste by rail from transfer stations.

Debate

Use of incinerators for waste management is controversial. The debate over incinerators typically involves business interests (representing both waste generators and incinerator firms), government regulators, environmental activists and local citizens who must weigh the economic appeal of local industrial activity with their concerns over health and environmental risk.

People and organizations professionally involved in this issue include the U.S. Environmental Protection Agency and a great many local and national air quality regulatory agencies worldwide.

Arguments for Incineration

Kehrichtverbrennungsanlage Zürcher Oberland (KEZO) in Hinwil, Switzerland

- The concerns over the health effects of dioxin and furan emissions have been significantly lessened by advances in emission control designs and very stringent new governmental regulations that have resulted in large reductions in the amount of dioxins and furans emissions.

- The U.K. Health Protection Agency concluded in 2009 that "Modern, well managed incinerators make only a small contribution to local concentrations of air pollutants. It is possible that such small additions could have an impact on health but such effects, if they exist, are likely to be very small and not detectable."

- Incineration plants can generate electricity and heat that can substitute power plants powered by other fuels at the regional electric and district heating grid, and steam supply for industrial customers. Incinerators and other waste-to-energy plants generate at least partially biomass-based renewable energy that offsets greenhouse gas pollution from coal-, oil- and gas-fired power plants. The E.U. considers energy generated from biogenic waste (waste with biological origin) by incinerators as non-fossil renewable energy under its emissions caps. These greenhouse gas reductions are in addition to those generated by the avoidance of landfill methane.

- The bottom ash residue remaining after combustion has been shown to be a non-hazardous solid waste that can be safely put into landfills or recycled as construction aggregate. Samples are tested for ecotoxic metals.

- In densely populated areas, finding space for additional landfills is becoming increasingly difficult.

The Maishima waste treatment center in Osaka, designed by Friedensreich Hundertwasser, uses heat for power generation.

Fine particles can be efficiently removed from the flue gases with baghouse filters. Even though approximately 40% of the incinerated waste in Denmark was incinerated at plants with no baghouse filters, estimates based on measurements by the Danish Environmental Research Institute showed that incinerators were only responsible for approximately 0.3% of the total domestic emissions of particulate smaller than 2.5 micrometres ($PM_{2.5}$) to the atmosphere in 2006.

- Incineration of municipal solid waste avoids the release of methane. Every ton of MSW incinerated, prevents about one ton of carbon dioxide equivalents from being released to the atmosphere.

- Most municipalities that operate incineration facilities have higher recycling rates than neighboring cities and counties that do not send their waste to incinerators. This is in part due to enhanced recovery of ceramic materials reused in construction, as well as ferrous and in some cases non-ferrous metals that can be recovered from combustion residue. Metals recovered from ash would typically be difficult or impossible to recycle through conventional means, as the removal of attached combustible material through incineration provides an alternative to labor- or energy-intensive mechanical separation methods.

- Volume of combusted waste is reduced by approximately 90%, increasing the life of landfills. Ash from modern incinerators is vitrified at temperatures of 1,000 °C (1,830 °F) to 1,100 °C (2,010 °F), reducing the leachability and toxicity of residue. As a result, special landfills are generally no longer required for incinerator ash from municipal waste streams, and existing landfills can see their life dramatically increased by combusting waste, reducing the need for municipalities to site and construct new landfills.

Arguments Against Incineration

Decommissioned Kwai Chung Incineration Plant from 1978. It was demolished by February 2009.

- The Scottish Protection Agency's (SEPA) comprehensive health effects research concluded "inconclusively" on health effects in October 2009. The authors stress, that even though no conclusive evidence of non-occupational health effects from incinerators were found in the existing literature, "small but important effects might be virtually impossible to detect". The report highlights epidemiological deficiencies in previous UK health studies and suggests areas for future studies. The U.K. Health Protection Agency produced a lesser summary in September 2009. Many toxiocologists criticise and dispute this report as not being comprehensive epidemiologically, thin on peer review and the effects of fine particle effects on health.

- The highly toxic fly ash must be safely disposed of. This usually involves additional waste miles and the need for specialist toxic waste landfill elsewhere. If not done properly, it may cause concerns for local residents.

- The health effects of dioxin and furan emissions from old incinerators; especially during start up and shut down, or where filter bypass is required continue to be a problem.

- Incinerators emit varying levels of heavy metals such as vanadium, manganese, chromium, nickel, arsenic, mercury, lead, and cadmium, which can be toxic at very minute levels.

- Incinerator Bottom Ash (IBA) has elevated levels of heavy metals with ecotoxicity concerns if not reused properly. Some people have the opinion that IBA reuse is still in its infancy and is still not considered to be a mature or desirable product, despite additional engineering treatments. Concerns of IBA use in Foam Concrete have been expressed by the UK Health and Safety Executive in 2010 following several construction and demolition explosions. In its guidance document, IBA is currently banned from use by the UK Highway Authority in concrete work until these incidents have been investigated.

- Alternative technologies are available or in development such as mechanical biological treatment, anaerobic digestion (MBT/AD), autoclaving or mechanical heat treatment (MHT) using steam or plasma arc gasification (PGP), which is incineration using electrically produced extreme high temperatures, or combinations of these treatments.

- Erection of incinerators compete with the development and introduction of other emerging technologies. A UK government WRAP report, August 2008 found that in the UK median incinerator costs per ton were generally higher than those for MBT treatments by £18 per metric ton; and £27 per metric ton for most modern (post 2000) incinerators.

- Building and operating waste processing plants such as incinerators requires long contract periods to recover initial investment costs, causing a long term lock-in. Incinerator lifetimes normally range 25–30 years. This was highlighted by Peter Jones, OBE, the Mayor of London's waste representative in April 2009.

- Incinerators produce fine particles in the furnace. Even with modern particle filtering of the flue gases, a small part of these is emitted to the atmosphere. $PM_{2.5}$ is not separately regulated in the European Waste Incineration Directive, even though they are repeatedly correlated spatially to infant mortality in the UK (M. Ryan's ONS data based maps around the EfW/CHP waste incinerators at Edmonton, Coventry, Chineham, Kirklees and Sheffield). Under WID there is no requirement to monitor stack top or downwind incinerator $PM_{2.5}$ levels. Several European doctors associations (including cross discipline experts such as physicians, environmental chemists and toxicologists) in June 2008 representing over 33,000 doctors wrote a keynote statement directly to the European Parliament citing widespread concerns on incinerator particle emissions and the absence of specific fine and ultrafine particle size monitoring or in depth industry/government epidemiological studies of these minute and invisible incinerator particle size emissions.

- Local communities are often opposed to the idea of locating waste processing plants such as incinerators in their vicinity (the Not in My Back Yard phenomenon). Studies in Andover, Massachusetts correlated 10% property devaluations with close incinerator proximity.

- Prevention, waste minimisation, reuse and recycling of waste should all be preferred to incineration according to the waste hierarchy. Supporters of zero waste consider incinerators and other waste treatment technologies as barriers to recycling and separation beyond particular levels, and that waste resources are sacrificed for energy production.

- A 2008 Eunomia report found that under some circumstances and assumptions, incineration causes less CO_2 reduction than other emerging EfW and CHP technology combinations for treating residual mixed waste. The authors found that CHP incinerator technology without waste recycling ranked 19 out of 24 combinations (where all alternatives to incineration were combined with advanced waste recycling plants); being 228% less efficient than the ranked 1 Advanced MBT maturation technology; or 211% less efficient than plasma gasification/autoclaving combination ranked 2.

- Some incinerators are visually undesirable. In many countries they require a visually intrusive chimney stack.

- If reusable waste fractions are handled in waste processing plants such as incinerators in developing nations, it would cut out viable work for local economies. It is estimated that there are 1 million people making a livelihood off collecting waste.

- The reduced levels of emissions from municipal waste incinerators and waste to energy plants from historical peaks are largely the product of the proficient use of emission control technology. Emission controls add to the initial and operational expenses. It should not be assumed that all new plants will employ the best available control technology if not required by law.

Trends in Incinerator Use

The history of municipal solid waste (MSW) incineration is linked intimately to the history of landfills and other waste treatment technology. The merits of incineration are inevitably judged in relation to the alternatives available. Since the 1970s, recycling and other prevention measures have changed the context for such judgements. Since the 1990s alternative waste treatment technologies have been maturing and becoming viable.

Incineration is a key process in the treatment of hazardous wastes and clinical wastes. It is often imperative that medical waste be subjected to the high temperatures of incineration to destroy pathogens and toxic contamination it contains.

Incineration in North America

The first incinerator in the U.S. was built in 1885 on Governors Island in New York. In 1949, Robert C. Ross founded one of the first hazardous waste management companies in the U.S. He began Robert Ross Industrial Disposal because he saw an opportunity to meet the hazardous waste management needs of companies in northern Ohio. In 1958, the company built one of the first hazardous waste incinerators in the U.S.

The first full-scale, municipally operated incineration facility in the U.S. was the Arnold O. Chantland Resource Recovery Plant, built in 1975 and located in Ames, Iowa. This plant is still in operation and produces refuse-derived fuel that is sent to local power plants for fuel. The first commercially successful incineration plant in the U.S. was built in Saugus, Massachusetts in October 1975 by Wheelabrator Technologies, and is still in operation today.

There are several environmental or waste management corporations that transport ultimately to an incinerator or cement kiln treatment center. Currently (2009), there are three main businesses that incinerate waste: Clean Harbours, WTI-Heritage, and Ross Incineration Services. Clean Harbours has acquired many of the smaller, independently run facilities, accumulating 5–7 incinerators in the process across the U.S. WTI-Heritage has one incinerator, located in the southeastern corner of Ohio across the Ohio River from West Virginia.

Several old generation incinerators have been closed; of the 186 MSW incinerators in 1990, only 89 remained by 2007, and of the 6200 medical waste incinerators in 1988, only 115 remained in 2003. No new incinerators were built between 1996 and 2007. The main reasons for lack of activity have been:

- Economics. With the increase in the number of large inexpensive regional landfills and, up until recently, the relatively low price of electricity, incinerators were not able to compete for the 'fuel', i.e., waste in the U.S.

- Tax policies. Tax credits for plants producing electricity from waste were rescinded in the U.S. between 1990 and 2004.

There has been renewed interest in incineration and other waste-to-energy technologies in the U.S. and Canada. In the U.S., incineration was granted qualification for renewable energy production tax credits in 2004. Projects to add capacity to existing plants are underway, and municipalities are once again evaluating the option of building incineration plants rather than continue landfilling municipal wastes. However, many of these projects have faced continued political opposition in spite of renewed arguments for the greenhouse gas benefits of incineration and improved air pollution control and ash recycling.

Incineration in Europe

In Europe, with the ban on landfilling untreated waste, scores of incinerators have been built in the last decade, with more under construction. Recently, a number of municipal governments have begun the process of contracting for the construction and operation of incinerators. In Europe, some of the electricity generated from waste is deemed to be from a 'Renewable Energy Source (RES) and is thus eligible for tax credits if privately operated. Also, some incinerators in Europe are equipped with waste recovery, allowing the reuse of ferrous and non-ferrous materials found in landfills. A prominent example is the AEB Waste Fired Power Plant.

In Sweden, about 50% of the generated waste is burned in waste-to-energy facilities, producing electricity and supplying local cities' district heating systems. The importance of waste in Sweden's electricity generation scheme is reflected on their 700.000 tons of waste imported per year to supply waste-to-energy facilities.

Incineration in The United Kingdom

The technology employed in the UK waste management industry has been greatly lagging behind that of Europe due to the wide availability of landfills. The Landfill Directive set down by the European Union led to the Government of the United Kingdom imposing waste legislation including the landfill tax and Landfill Allowance Trading Scheme. This legislation is designed to reduce the release of greenhouse gases produced by landfills through the use of alternative methods of waste treatment. It is the UK Government's position that incineration will play an increasingly large role in the treatment of municipal waste and supply of energy in the UK.

In 2008, plans for potential incinerator locations exists for approximately 100 sites. These have been interactively mapped by UK NGO's.

Under a new plan in June 2012, a DEFRA-backed grant scheme (The Farming and Forestry Improvement Scheme) was set up to encourage the use of low-capacity incinerators on agricultural sites to improve their bio security.

Incineration Units for Emergency Use

Emergency incineration systems exist for the urgent and biosecure disposal of animals and their by-products following a mass mortality or disease outbreak. An increase in regulation and enforcement from governments and institutions worldwide has been forced through public pressure and significant economic exposure.

Mobile incineration unit for emergency use

Contagious animal disease has cost governments and industry $200 billion over 20 years to 2012 and is responsible for over 65% of infectious disease outbreaks worldwide in the past sixty years. One-third of global meat exports (approx 6 million tonnes) is affected by trade restrictions at any time and as such the focus of Governments, public bodies and commercial operators is on cleaner, safer and more robust methods of animal carcass disposal to contain and control disease.

Large-scale incineration systems are available from niche suppliers and are often bought by governments as a safety net in case of contagious outbreak. Many are mobile and can be quickly deployed to locations requiring biosecure disposal.

Small Incinerator Units

Small-scale incinerators exist for special purposes. For example, the small-scale incinerators are aimed for hygienically safe destruction of medical waste in developing countries. Small incinerators can be quickly deployed to remote areas where an outbreak has occurred to dispose of infected animals quickly and without the risk of cross contamination.

An example of a low capacity, mobile incinerator

In Popular Media

- In Cube Zero, fictional so-called "flash Incinerators" exist, which essentially vaporize anything organic.

- Incinerators make an appearance in SimCity 3000 in two varieties: a large, traditional combustion device that spews out a significant amount of air pollution, and a more modern device that converts the waste into energy to power the city with a bigger capacity to load the garbage, though still producing a lot of pollution.

- They also make an appearance in SimCity 4, but without the non-energy-from-waste variant.

- In the climax of Portal (video game), the main protagonist, Chell, while on a conveyor belt, escapes an incinerator, after the game's main antagonist, GLaDOS, forced her into it.

- The climax of Toy Story 3 features an infamous scene, where the working of a moving-grate incinerator (and of a garbage shredder) was shown dramatically from the inside, as the toys face destruction.

Pyrolysis

Pyrolysis is a thermochemical decomposition of organic material at elevated temperatures in the absence of oxygen (or any halogen). It involves the simultaneous change of chemical composition and physical phase, and is irreversible. The word is coined from the Greek-derived elements *pyro* "fire" and *lysis* "separating".

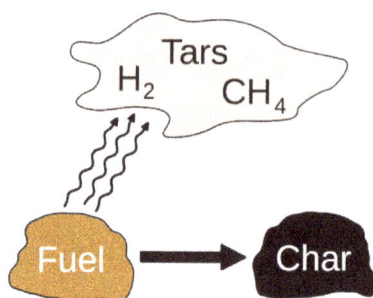

Simplified depiction of pyrolysis chemistry.

Pyrolysis is a type of thermolysis, and is most commonly observed in organic materials exposed to high temperatures. It is one of the processes involved in charring wood, starting at 200–300 °C (390–570 °F). It also occurs in fires where solid fuels are burning or when vegetation comes into contact with lava in volcanic eruptions. In general, pyrolysis of organic substances produces gas and liquid products and leaves a solid residue richer in carbon content, char. Extreme pyrolysis, which leaves mostly carbon as the residue, is called carbonization.

The process is used heavily in the chemical industry, for example, to produce charcoal, activated carbon, methanol, and other chemicals from wood, to convert ethylene dichloride into vinyl chloride to make PVC, to produce coke from coal, to convert biomass into syngas and biochar, to turn waste plastics back into usable oil, or waste into safely disposable substances, and for transforming medium-weight hydrocarbons from oil into lighter ones like gasoline. These specialized uses of pyrolysis may be called various names, such as dry distillation, destructive distillation, or cracking. Pyrolysis is also used in the creation of nanoparticles, zirconia and oxides utilizing an ultrasonic nozzle in a process called ultrasonic spray pyrolysis (USP).

Pyrolysis also plays an important role in several cooking procedures, such as baking, frying, grilling, and caramelizing. It is a tool of chemical analysis, for example, in mass spectrometry and in carbon-14 dating. Indeed, many important chemical substances, such as phosphorus and sulfuric acid, were first obtained by this process. Pyrolysis has been assumed to take place during catagenesis, the conversion of buried organic matter to fossil fuels. It is also the basis of pyrography. In their embalming process, the ancient Egyptians used a mixture of substances, including methanol, which they obtained from the pyrolysis of wood.

Pyrolysis differs from other processes like combustion and hydrolysis in that it usually does not involve reactions with oxygen, water, or any other reagents. In practice, it is not possible to achieve a completely oxygen-free atmosphere. Because some oxygen is present in any pyrolysis system, a small amount of oxidation occurs.

The term has also been applied to the decomposition of organic material in the presence of superheated water or steam (hydrous pyrolysis), for example, in the steam cracking of oil.

Occurrence and Uses

Fire

Pyrolysis is usually the first chemical reaction that occurs in the burning of many solid organic fuels, like wood, cloth, and paper, and also of some kinds of plastic. In a wood fire, the visible flames are not due to combustion of the wood itself, but rather of the gases released by its pyrolysis, whereas the flame-less burning of a solid, called smouldering, is the combustion of the solid residue (char or charcoal) left behind by pyrolysis. Thus, the pyrolysis of common materials like wood, plastic, and clothing is extremely important for fire safety and firefighting. In pyrolysis there is a gas phase present. It should not be confused with hydrothermal reactions such as hydrothermal gasification, hydrothermal liquidation, and hydrothermal carbonization, which occur in aqueous environments because the temperatures and reaction pathways differ, with ionic reactions favored in aqueous reactions and radical reactions favored in the absence of water.

Cooking

Pyrolysis occurs whenever food is exposed to high enough temperatures in a dry environment, such as roasting, baking, toasting, or grilling. It is the chemical process responsible for the formation of the golden-brown crust in foods prepared by those methods.

In normal cooking, the main food components that undergo pyrolysis are carbohydrates (including sugars, starch, and fibre) and proteins. Pyrolysis of fats requires a much higher temperature, and, since it produces toxic and flammable products (such as acrolein), it is, in general, avoided in normal cooking. It may occur, however, when grilling fatty meats over hot coals.

Even though cooking is normally carried out in air, the temperatures and environmental conditions are such that there is little or no combustion of the original substances or their decomposition products. In particular, the pyrolysis of proteins and carbohydrates begins at temperatures much lower than the ignition temperature of the solid residue, and the volatile subproducts are too diluted in air to ignite. (In flambé dishes, the flame is due mostly to combustion of the alcohol, while the crust is formed by pyrolysis as in baking.)

Pyrolysis of carbohydrates and proteins requires temperatures substantially higher than 100 °C (212 °F), so pyrolysis does not occur as long as free water is present, e.g., in boiling food — not even in a pressure cooker. When heated in the presence of water, carbohydrates and proteins suffer gradual hydrolysis rather than pyrolysis. Indeed, for most foods, pyrolysis is usually confined to the outer layers of food, and begins only after those layers have dried out.

Food pyrolysis temperatures are, however, lower than the boiling point of lipids, so pyrolysis occurs when frying in vegetable oil or suet, or basting meat in its own fat.

Pyrolysis also plays an essential role in the production of barley tea, coffee, and roasted nuts such as peanuts and almonds. As these consist mostly of dry materials, the process of pyrolysis is not limited to the outermost layers but extends throughout the materials. In all these cases, pyrolysis creates or releases many of the substances that contribute to the flavor, color, and biological properties of the final product. It may also destroy some substances that are toxic, unpleasant in taste, or those that may contribute to spoilage.

Controlled pyrolysis of sugars starting at 170 °C (338 °F) produces caramel, a beige to brown water-soluble product widely used in confectionery and (in the form of caramel coloring) as a coloring agent for soft drinks and other industrialized food products.

Solid residue from the pyrolysis of spilled and splattered food creates the brown-black encrustation often seen on cooking vessels, stove tops, and the interior surfaces of ovens.

Charcoal

People have used pyrolysis for turning wood into charcoal on an industrial scale since ancient times. Besides wood, the process can also use sawdust and other wood-waste products.

Charcoal is obtained by heating wood until its complete pyrolysis (carbonization) occurs, leaving only carbon and inorganic ash. In many parts of the world charcoal is still produced semi-industrially by burning a pile of wood that has been mostly covered with mud or with bricks. The heat generated by burning part of the wood and the volatile byproducts pyrolyzes the rest of the pile. The limited supply of oxygen prevents the charcoal from burning. A more modern alternative is to heat the wood in an airtight metal vessel, which is much less polluting and allows the volatile products to be condensed.

The original vascular structure of the wood and the pores created by escaping gases combine to produce a light and porous material. By starting with a dense wood-like material, such as nutshells or peach stones, one obtains a form of charcoal with particularly fine pores (and hence a much larger pore surface area), called activated carbon, which is used as an adsorbent for a wide range of chemical substances.

Biochar

Residues of incomplete organic pyrolysis, e.g., from cooking fires, are thought to be the key component of the terra preta soils associated with ancient indigenous communities of the Amazon basin. Terra preta is much sought by local farmers for its superior fertility compared to the natural red soil of the region. Efforts are underway to recreate these soils through biochar, the solid residue of pyrolysis of various materials, mostly organic waste.

Biochar improves the soil texture and ecology, increasing its ability to retain fertilizers and release them slowly. It naturally contains many of the micronutrients needed by plants, such as selenium. It is also safer than other "natural" fertilizers such as animal manure, since it has been disinfected at high temperature. And, since it releases its nutrients at a slow rate, it greatly reduces the risk of water table contamination.

Biochar is also being considered for carbon sequestration, with the aim of mitigation of global warming. The solid, carbon-containing char produced can be sequestered in the ground, where it will remain for several hundred to a few thousand years.

Coke

Pyrolysis is used on a massive scale to turn coal into coke for metallurgy, especially steelmaking. Coke can also be produced from the solid residue left from petroleum refining.

Those starting materials typically contain hydrogen, nitrogen, or oxygen atoms combined with carbon into molecules of medium to high molecular weight. The coke-making or "coking" process consists of heating the material in closed vessels to very high temperatures (up to 2,000 °C or 3,600 °F) so that those molecules are broken down into lighter volatile substances, which leave the vessel, and a porous but hard residue that is mostly carbon and inorganic ash. The amount of volatiles varies with the source material, but is typically 25–30% of it by weight.

Carbon Fiber

Carbon fibers are filaments of carbon that can be used to make very strong yarns and textiles. Carbon fiber items are often produced by spinning and weaving the desired item from fibers of a suitable polymer, and then pyrolyzing the material at a high temperature (from 1,500–3,000 °C or 2,730–5,430 °F).

The first carbon fibers were made from rayon, but polyacrylonitrile has become the most common starting material.

For their first workable electric lamps, Joseph Wilson Swan and Thomas Edison used carbon filaments made by pyrolysis of cotton yarns and bamboo splinters, respectively.

Pyrolytic Carbon

Pyrolysis is the reaction used to coat a preformed substrate with a layer of pyrolytic carbon. This is typically done in a fluidized bed reactor heated to 1,000–2,000 °C or 1,830–3,630 °F. Pyrolytic carbon coatings are used in many applications, including artificial heart valves.

Biofuel

Pyrolysis is the basis of several methods that are being developed for producing fuel from biomass, which may include either crops grown for the purpose or biological waste products from other industries. Crops studied as biomass feedstock for pyrolysis include native North American prairie grasses such as *switchgrass* and bred versions of other grasses such as *Miscantheus giganteus*. Crops and plant material wastes provide biomass feedstock on the basis of their lignocellulose portions.

Although synthetic diesel fuel cannot yet be produced directly by pyrolysis of organic materials, there is a way to produce similar liquid (bio-oil) that can be used as a fuel, after the removal of valuable bio-chemicals that can be used as food additives or pharmaceuticals. Higher efficiency is achieved by flash pyrolysis, in which finely divided feedstock is quickly heated to between 350 and 500 °C (660 and 930 °F) for less than 2 seconds.

Fuel bio-oil can also be produced by hydrous pyrolysis from many kinds of feedstock, including waste from pig and turkey farming, by a process called thermal depolymerization (which may, however, include other reactions besides pyrolysis).

Adhesives

Neanderthals used pyrolysis of birch bark to produce a pitch with which they secured flaked stones to spear shafts. Recently, researchers have developed a process to pyrolyze birch bark to produce an oil that can replace phenol in phenol formaldehyde resin (these resins are mostly used to manufacture plywood).

Pesticides

Pyrolysis can also be used to produce pesticides from biomass.

Plastic Waste Disposal

Anhydrous pyrolysis can also be used to produce liquid fuel similar to diesel from plastic waste, with a higher cetane value and lower sulphur content than traditional diesel. Using pyrolysis to extract fuel from end-of-life plastic is a second-best option after recycling, is environmentally preferable to landfill, and can help reduce dependency on foreign fossil fuels and geo-extraction.

Waste Tire Disposal

In the United States alone, over 290 million car tires are discarded annually. Pyrolysis of scrap or waste tires (WT) is an attractive alternative to disposal in landfills, allowing the high energy content of the tire to be recovered as fuel. Using tires as fuel produce equal energy as burning oil and 25% more energy than burning coal.

An average car tire is made up of 50-60% hydrocarbons, resulting in a yield of 38-56% oil, 10-30% gas and 14-56% char. The oil produced is largely composed of benzene, diesel, kerosene, fuel oil and heavy fuel oil, while the produced gas has a similar composition to natural gas. The proportion and the purity of the products are governed by two major factors:

1. Environment (e.g. pressure, temperature, time, reactor type)

2. Material (e.g. age, composition, size, type)

As car tires age, they increase in hardness, making it more difficult for pyrolysis to break the molecules into shorter chains. This shifts the yield composition towards diesel oil which is composed of larger molecules. Conversely, an increase in temperature increases the likelihood of breaking the molecule chain and shifts the yield composition towards benzene oil which is composed of smaller molecules. Other products from car tire pyrolysis include steel wires, carbon black and bitumen.

Although the pyrolysis of WT has been widely developed throughout the world, there are legislative, economic, and marketing obstacles to widespread adoption. Oil derived from tire rubber pyrolysis contains high sulfur content, which gives it high potential as a pollutant and should be desulfurized A number of prototype and full-scale pyrolysis plants specialized in carbon black production have successfully established across the world, including the United States, France, Germany and Japan. Because carbon black is used for pigment, rubber strengthening and UV protection, it is a relatively large and growing market. Pyrolysis plants specialized in fuel oil production is not an implausible concept. However, as profits of such ventures come from the added value between the production and distillation of oil, there is little profit without vertical integration in the oil industry. The inconsistency of the feedstock makes it very difficult to control the uniformity of the products and makes oil companies hesitant to purchase oil produced via pyrolysis. Finally, the cost of producing oil through conventional means is generally less expensive than this alternative. To date, there is no known commercially profitable standalone pyrolysis plant that specializes in oil production. However, with funding to upgrade pyrolysis oil to light fuel grade, this may be possible. Nevertheless, pyrolysis is a valuable method for disposing waste tires.

Chemical Analysis

Pyrolysis can be used for the molecular characterisation of molecules when used in conjunction with gas chromatography-mass spectrometry (Py-GC-MS). This technique has been used to analyse the method and products of fungal decay of wood.

Thermal Cleaning

Pyrolysis is also used for *thermal cleaning*, an industrial application to remove organic substances such as polymers, plastics and coatings from parts, products or production components like extruder screws, spinnerets and static mixers. During the thermal cleaning process, at temperatures between 600 °F to 1000 °F (310 C° to 540 C°), organic material is converted by pyrolysis and oxidation into volatile organic compounds, hydrocarbons and carbonized gas. Inorganic elements remain.

Several types of thermal cleaning systems use pyrolysis:

- *Molten Salt Baths* belong to the oldest thermal cleaning systems; cleaning with a molten salt bath is very fast but implies the risk of dangerous splatters, or other potential hazards connected with the use of salt baths, like explosions or highly toxic hydrogen cyanide gas;

- *Fluidized Bed Systems* use sand or aluminium oxide as heating medium; these systems also clean very fast but the medium does not melt or boil, nor emit any vapors or odors; the cleaning process takes one to two hours;

- *Vacuum Ovens* use pyrolysis in a vacuum avoiding uncontrolled combustion inside the cleaning chamber; the cleaning process takes 8 to 30 hours;

- *Burn-Off Ovens*, also known as *Heat-Cleaning Ovens*, are gas-fired and used in the painting, coatings, electric motors and plastics industries for removing organics from heavy and large metal parts.

Processes

In many industrial applications, the process is done under pressure and at operating temperatures above 430 °C (806 °F). For agricultural waste, for example, typical temperatures are 450 to 550 °C (840 to 1,000 °F).

Processes

Since pyrolysis is endothermic, various methods to provide heat to the reacting biomass particles have been proposed:

- Partial combustion of the biomass products through air injection. This results in poor-quality products.

- Direct heat transfer with a hot gas, the ideal one being product gas that is reheated and recycled. The problem is to provide enough heat with reasonable gas flow-rates.

- Indirect heat transfer with exchange surfaces (wall, tubes): it is difficult to achieve good heat transfer on both sides of the heat exchange surface.

- Direct heat transfer with circulating solids: solids transfer heat between a burner and a pyrolysis reactor. This is an effective but complex technology.

For flash pyrolysis, the biomass must be ground into fine particles and the insulating char layer that forms at the surface of the reacting particles must be continuously removed. The following technologies have been proposed for biomass pyrolysis:

- Fixed beds used for the traditional production of charcoal: poor, slow heat transfer result in very low liquid yields.

- Augers: this technology is adapted from a Lurgi process for coal gasification. Hot sand and biomass particles are fed at one end of a screw. The screw mixes the sand and biomass and conveys them along. It provides a good control of the biomass residence time. It does not dilute the pyrolysis products with a carrier or fluidizing gas. However, sand must be reheated in a separate vessel, and mechanical reliability is a concern. There is no large-scale commercial implementation.

- Electrically heated augers: one process uses an electrical current passed through an auger to heat the material giving excellent heat transfer by contact and radiation to the waste material.

- Ablative processes: biomass particles are moved at high speed against a hot metal surface. Ablation of any char forming at a particle's surface maintains a high rate of heat transfer. This can be achieved by using a metal surface spinning at high speed within a bed of biomass particles, which may present mechanical reliability problems but prevents any dilution of the products. As an alternative, the particles may be suspended in a carrier gas and introduced at high speed through a cyclone whose wall is heated; the products are diluted with the carrier gas. A problem shared with all ablative processes is that scale-up is made difficult, since the ratio of the wall surface to the reactor volume decreases as the reactor size is increased. There is no large-scale commercial implementation.

- Rotating cone: pre-heated hot sand and biomass particles are introduced into a rotating cone. Due to the rotation of the cone, the mixture of sand and biomass is transported across the cone surface by centrifugal force. The process is offered by BTG-BTL, a subsidiary from BTG Biomass Technology Group B.V. in The Netherlands. Like other shallow transported-bed reactors relatively fine particles (several mm) are required to obtain a liquid yield of around 70 wt.%. Larger-scale commercial implementation (up to 5 t/h input) is underway.

- Fluidized beds: biomass particles are introduced into a bed of hot sand fluidized by a gas, which is usually a recirculated product gas. High heat transfer rates from fluidized sand result in rapid heating of biomass particles. There is some ablation by attrition with the sand particles, but it is not as effective as in the ablative processes. Heat is usually provided by heat exchanger tubes through which hot combustion gas flows. There is some dilution of the products, which makes it more difficult to condense and then remove the bio-oil mist from the gas exiting the condensers. This process has been scaled up by companies such as Dynamotive and Agri-Therm. The main challenges are in improving the quality and consistency of the bio-oil.

- Circulating fluidized beds: biomass particles are introduced into a circulating fluidized bed of hot sand. Gas, sand, and biomass particles move together, with the transport gas usually being a recirculated product gas, although it may also be a combustion gas. High heat transfer rates from sand ensure rapid heating of biomass particles and ablation stronger than with regular fluidized beds. A fast separator separates the product gases and vapors from the sand and char particles. The sand particles are reheated in a fluidized burner vessel and recycled to the reactor. Although this process can be easily scaled up, it is rather complex and the products are much diluted, which greatly complicates the recovery of the liquid products.

- Mechanical Fluidized Reactor (MFR). A mechanical stirrer agitates a hot bed of pure char particles into which biomass particles are injected. The stirrer also enhances heat transfer from the reactor wall to the agitated bed. No fluidization gas is required: evolving vapors aerate the bed and greatly reduce the power consumption of the mechanical stirrer. This compact reactor has been used for a mobile pyrolysis plant.

- Chain grate: dry biomass is fed onto a hot (500 °C) heavy cast metal grate or apron which forms a continuous loop. A small amount of air aids in heat transfer and in primary reactions for drying and carbonization. Volatile products are combusted for process and boiler heating.

Use of Vacuum

In vacuum pyrolysis, organic material is heated in a vacuum to decrease its boiling point and avoid adverse chemical reactions. Called flash vacuum pyrolysis, this approach is used in organic synthesis.

Industrial Sources

Many sources of organic matter can be used as feedstock for pyrolysis. Suitable plant material includes greenwaste, sawdust, waste wood, woody weeds; and agricultural sources including nut shells, straw, cotton trash, rice hulls, switch grass; and animal waste including poultry litter, dairy manure, and potentially other manures. Pyrolysis is used as a form of thermal treatment to reduce waste volumes of domestic refuse. Some industrial byproducts are also suitable feedstock including paper sludge and distillers grain.

There is also the possibility of integrating with other processes such as mechanical bio-logical treatment and anaerobic digestion.

Industrial Products

- syngas (flammable mixture of carbon monoxide and hydrogen): can be produced in sufficient quantities to provide both the energy needed for pyrolysis and some excess production

- solid char that can either be burned for energy or be recycled as a fertilizer (bio-char).

Fire Protection

Destructive fires in buildings will often burn with limited oxygen supply, resulting in pyrolysis reactions. Thus, pyrolysis reaction mechanisms and the pyrolysis properties of materials are important in fire protection engineering for passive fire protection. Pyrolytic carbon is also important to fire investigators as a tool for discovering origin and cause of fires.

Chemistry

Current research examines the multiple reaction pathways of pyrolysis to understand how to manipulate the formation of pyrolysis' multiple products (oil, gas, char, and miscellaneous chemicals) to enhance the economic value of pyrolysis; identifying catalysts to manipulate pyrolysis reactions is also a goal of some pyrolysis research. Published research suggests that pyrolysis reactions have some dependence upon the structural composition of feedstocks (e.g. lignocellulosic biomass), with contributions from some minerals present in the feedstocks; some minerals present in feedstock are thought to increase the cost of operation of pyrolysis or decrease the value of oil produced from pyrolysis, through corrosive reactions. The low quality of oils produced through pyrolysis can be improved by subjecting the oils to one or many physical and chemical processes, which might drive production costs, but may make sense economically as circumstances change.

Landfill

A landfill site (also known as a tip, dump, rubbish dump, garbage dump or dumping ground and historically as a midden) is a site for the disposal of waste materials by burial and is the oldest form of waste treatment (although the burial part is modern; historically, refuse was just left in piles or thrown into pits). Historically, landfills have been the most common method of organized waste disposal and remain so in many places around the world.

Some landfills are also used for waste management purposes, such as the temporary storage, consolidation and transfer, or processing of waste material (sorting, treatment, or recycling).

A landfill in Poland

A landfill also may refer to ground that has been filled in with rocks instead of waste materials, so that it can be used for a specific purpose, such as for building houses. Unless they are stabilized, these areas may experience severe shaking or soil liquefaction of the ground during a large earthquake.

Operations

Typically, operators of well-run landfills for non-hazardous waste meet predefined specifications by applying techniques to:

One of several landfills used by Dryden, Ontario, Canada.

1. confine waste to as small an area as possible

2. compact waste to reduce volume

3. cover waste (usually daily) with layers of soil

During landfill operations a scale or weighbridge may weigh waste-collection vehicles on arrival and personnel may inspect loads for wastes that do not accord with the landfill's waste-acceptance criteria. Afterward, the waste-collection vehicles use the existing road network on their way to the tipping face or working front, where they unload their contents. After loads are deposited, compactors or bulldozers can spread and compact the waste on the working face. Before leaving the landfill boundaries, the waste collection vehicles may pass through a wheel-cleaning facility. If necessary, they return to the weighbridge for re-weighing without their load. The weighing process can assemble statistics on the daily incoming waste-tonnage, which databases can retain for record keeping. In addition to trucks, some landfills may have equipment to handle railroad containers. The use of "rail-haul" permits landfills to be located at more remote sites, without the problems associated with many truck trips.

Typically, in the working face, the compacted waste is covered with soil or alternative materials daily. Alternative waste-cover materials include chipped wood or other "green waste", several sprayed-on foam products, chemically "fixed" bio-solids, and temporary blankets. Blankets can be lifted into place at night and then removed the following day prior to waste placement. The space that is occupied daily by the compacted waste and the cover material is called a daily cell. Waste compaction is critical to extending the life of the landfill. Factors such as waste compressibility, waste-layer thickness and the number of passes of the compactor over the waste affect the waste densities.

Advantages

Landfills are often the most cost-efficient way to dispose of waste, especially in countries like the United States with large open spaces. While resource recovery and incineration both require extensive investments in infrastructure, and material recovery also requires extensive manpower to maintain, landfills have fewer fixed—or ongoing—costs, allowing them to compete favorably. In addition, landfill gas can be upgraded to natural gas—landfill gas utilization—which is a potential revenue stream.

Social and Environmental Impact

Landfills have the potential to cause a number of issues. Infrastructure disruption, such as damage to access roads by heavy vehicles, may occur. Pollution of local roads and water courses from wheels on vehicles when they leave the landfill can be significant and can be mitigated by wheel washing systems. Pollution of the local environment, such as contamination of groundwater or aquifers or soil contamination may occur, as well.

Landfill operation in Hawaii. Note that the area being filled is a single, well-defined "cell" and that a protective landfill liner is in place (exposed on the left) to prevent contamination by leachates migrating downward through the underlying geological formation.

Leachate

Extensive efforts are made to capture and treat leachate from landfills before it reaches groundwater aquifers, but engineered liners always have a lifespan, though it may be 100 years or more. Eventually, every landfill liner will leak, allowing the leachate to contaminate the groundwater. Installation of composite liners with flexible membrane and soil barrier is enforced by the EPA to ensure that leachate is withheld.

Dangerous Gases

Methane is naturally generated by decaying organic wastes in a landfill. It is a potent greenhouse gas, and can itself be a danger because it is flammable and potentially explosive. In properly managed landfills, gas is collected and utilized. This could range from simple flaring to landfill gas utilization.

Infections

Poorly run landfills may become nuisances because of vectors such as rats and flies which can cause infectious diseases. The occurrence of such vectors can be mitigated through the use of daily cover.

Other potential issues include wildlife disruption, dust, odor, noise pollution, and reduced local property values.

Landfill Gas

Gases are produced in landfills due to the anaerobic digestion by microbes. In a properly managed landfill this gas is collected and used. Its uses range from simple flaring to the landfill gas utilization and generation of electricity. Landfill gas monitoring alerts workers to the presence of a build-up of gases to a harmful level. In some countries, landfill gas recovery is extensive; in the United States, for example, more than 850 landfills have active landfill gas recovery systems.

Regional Practice

A landfill in Perth, Western Australia

South East New Territories Landfill, Hong Kong

Canada

Landfills in Canada are regulated by provincial environmental agencies and environmental protection acts (EPA). Older facilities tend to fall under current standards and are monitored for leaching. Some former locations have been converted to parkland.

European Union

In the European Union, individual states are obliged to enact legislation to comply with the requirements and obligations of the European Landfill Directive. In the UK this is the Waste Implementation Programme.

United Kingdom

Landfilling practices in the UK have had to change in recent years to meet the challenges of the European Landfill Directive. The UK now imposes landfill tax upon biodegradable waste which is put into landfills. In addition to this the Landfill Allowance Trading Scheme has been established for local authorities to trade landfill quotas in England. A different system operates in Wales where authorities are not able to 'trade' between themselves, but have allowances known as the Landfill Allowance Scheme.

United States

U.S. landfills are regulated by each state's environmental agency, which establishes minimum guidelines; however, none of these standards may fall below those set by the United States Environmental Protection Agency (EPA).

Permitting a landfill generally takes between 5 and 7 years, costs millions of dollars and requires rigorous siting, engineering and environmental studies and demonstrations to ensure local environmental and safety concerns are satisfied.

Microbial Topics

The status of a landfill's microbial community may determine its digestive efficiency.

Bacteria that digest plastic have been found in landfills.

Reclaiming materials

Landfills can be regarded as a viable and abundant source of materials and energy. In the developing world, waste pickers often scavenge for still-usable materials. In a commercial context, landfill sites have also been discovered by companies, and many have begun harvesting materials and energy . Well known examples are gas recovery facilities. Other commercial facilities include waste incinerators which have built-in material recovery. This material recovery is possible through the use of filters (electro filter, active carbon and potassium filter, quench, HCl-washer, SO_2-washer, bottom ash-grating, etc.).

Alternatives

In addition to waste reduction and recycling strategies, there are various alternatives to landfills, including Waste-to-energy incineration, anaerobic digestion, composting, mechanical biological treatment, pyrolysis and plasma arc gasification, which have all begun to establish themselves in the market. Depending on local economics and incentives, these can be made more financially attractive than landfills.

Restrictions

Countries including Germany, Austria, Sweden, Denmark, Belgium, the Netherlands, and Switzerland, have banned the disposal of untreated waste in landfills. In these countries, only the ashes from incineration or the stabilized output of mechanical biological treatment plants may still be deposited.

References

- Hickmann, H. Lanier, Jr. (2003). American alchemy: the history of solid waste management in the United States. ForesterPress. ISBN 978-0-9707687-2-8.

- Ewald Schwing; Horst Uhrner (7 October 1999). "Method for removing polymer deposits which have formed on metal or ceramic machine parts, equipment and tools". Espacenet. European Patent Office. Retrieved 19 April 2016.

- "A Look at Thermal Cleaning Technology". ThermalProcessing.org. Process Examiner. 14 March 2014. Retrieved 4 December 2015.

- Gary Davis & Keith Brown (April 1996). "Cleaning Metal Parts and Tooling" (PDF). Pollution Prevention Regional Information Center. Process Heating. Retrieved 4 December 2015.

- Thomas S. Dwan (2 September 1980). "Process for vacuum pyrolysis removal of polymers from various objects". Espacenet. European Patent Office. Retrieved 26 December 2015.

- "Paint Stripping: Reducing Waste and Hazardous Material". Minnesota Technical Assistance Program. University of Minnesota. July 2008. Retrieved 4 December 2015.

- US EPA. "Tire-Derived Fuel - Scrap Tires". Archived from the original on March 26, 2015. Retrieved 18 February 2015.

- Refining fast pyrolysis of biomass. Thermo-Chemical Conversion of Biomass (Thesis). University of Twente. 2011. Retrieved 2012-05-30.

Radioactive Waste and its Management

Radioactive waste contains radioactive material which is a standard outcome of nuclear power generation technologies. One of the distinctive features of radioactive waste is that it decays over a period of time; hence it needs to be isolated and managed in an appropriate place for an adequate period of time until it no longer poses a threat. This chapter provides a comprehensive overview of the subject.

Radioactive Waste

Radioactive waste is waste that contains radioactive material. Radioactive waste is usually a by-product of nuclear power generation and other applications of nuclear fission or nuclear technology, such as research and medicine. Radioactive waste is hazardous to most forms of life and the environment, and is regulated by government agencies in order to protect human health and the environment.

Radioactivity naturally decays over time, so radioactive waste has to be isolated and confined in appropriate disposal facilities for a sufficient period until it no longer poses a threat. The time radioactive waste must be stored for depends on the type of waste and radioactive isotopes. It can range from a few days for highly radioactive isotopes to millions of years for slightly radioactive ones. Current major approaches to managing radioactive waste have been segregation and storage for short-lived waste, near-surface disposal for low and some intermediate level waste, and deep burial or partitioning / transmutation for the high-level waste.

A summary of the amounts of radioactive waste and management approaches for most developed countries are presented and reviewed periodically as part of the International Atomic Energy Agency (IAEA) Joint Convention on the Safety of Spent Fuel Management and on the Safety of Radioactive Waste Management.

Nature and Significance of Radioactive Waste

Radioactive waste typically comprises a number of radionuclides: unstable configurations of elements that decay, emitting ionizing radiation which can be harmful to humans and the environment. Those isotopes emit different types and levels of radiation, which last for different periods of time.

Physics

The radioactivity of all radioactive waste diminishes with time. All radionuclides contained in the waste have a half-life—the time it takes for half of the atoms to decay into another nuclide—and eventually all radioactive waste decays into non-radioactive elements (i.e., stable nuclides). Certain radioactive elements (such as plutonium-239) will remain hazardous to humans and other creatures for hundreds or thousands of years. Other radionuclides remain radioactive for millions of years (though most of these products have so little activity as a result of their long half-lives that their radiation is lost in the background level). Thus, these wastes must be shielded for centuries and isolated from the living environment for millennia. Since radioactive decay follows the half-life rule, the rate of decay is inversely proportional to the duration of decay. In other words, the radiation from a long-lived isotope like iodine-129 will be much less intense than that of a short-lived isotope like iodine-131. The two tables show some of the major radioisotopes, their half-lives, and their radiation yield as a proportion of the yield of fission of uranium-235.

The energy and the type of the ionizing radiation emitted by a radioactive substance are also important factors in determining its threat to humans. The chemical properties of the radioactive element will determine how mobile the substance is and how likely it is to spread into the environment and contaminate humans. This is further complicated by the fact that many radioisotopes do not decay immediately to a stable state but rather to radioactive decay products within a decay chain before ultimately reaching a stable state.

Pharmacokinetics

Exposure to radioactive waste may cause serious harm or death. In humans, a dose of 1 sievert carries a 5.5% risk of developing cancer, and this risk is assumed to be linearly proportional to dose even for low doses. Ionizing radiation causes deletions in chromosomes. If a developing organism such as an unborn child is irradiated, it is possible a birth defect may be induced, but it is unlikely this defect will be in a gamete or a gamete-forming cell. The incidence of radiation-induced mutations in humans is small, as in most mammals, because of natural cellular-repair mechanisms, many just now coming to light. These mechanisms range from DNA, mRNA and protein repair, to internal lysosomic digestion of defective proteins, and even induced cell suicide—apoptosis

Depending on the decay mode and the pharmacokinetics of an element (how the body processes it and how quickly), the threat due to exposure to a given activity of a radioisotope will differ. For instance iodine-131 is a short-lived beta and gamma emitter, but because it concentrates in the thyroid gland, it is more able to cause injury than caesium-137 which, being water soluble, is rapidly excreted in urine. In a similar way, the alpha emitting actinides and radium are considered very harmful as they tend to have long biological half-lives and their radiation has a high relative biological effectiveness, making it far more damaging to tissues per amount of energy deposited. Because of such differences, the rules determining biological injury differ widely according to the radioisotope, time of exposure and sometimes also the nature of the chemical compound which contains the radioisotope.

Sources of Waste

Radioactive waste comes from a number of sources. In countries with nuclear power plants, nuclear armament, or nuclear fuel treatment plants, the majority of waste originates from the nuclear fuel cycle and nuclear weapons reprocessing, otherwise there is no waste of nuclear origin. Other sources include medical and industrial wastes, as well as naturally occurring radioactive materials (NORM) that can be concentrated as a result of the processing or consumption of coal, oil and gas, and some minerals, as discussed below.

Nuclear Fuel Cycle

Front End

Waste from the front end of the nuclear fuel cycle is usually alpha-emitting waste from the extraction of uranium. It often contains radium and its decay products.

Uranium dioxide (UO_2) concentrate from mining is not very radioactive – only a thousand or so times as radioactive as the granite used in buildings. It is refined from yellowcake (U_3O_8), then converted to uranium hexafluoride gas (UF_6). As a gas, it undergoes enrichment to increase the U-235 content from 0.7% to about 4.4% (LEU). It is then turned into a hard ceramic oxide (UO_2) for assembly as reactor fuel elements.

The main by-product of enrichment is depleted uranium (DU), principally the U-238 isotope, with a U-235 content of ~0.3%. It is stored, either as UF_6 or as U_3O_8. Some is used in applications where its extremely high density makes it valuable such as anti-tank shells, even sailboat keels on at least one occasion. It is also used with plutonium for making mixed oxide fuel (MOX) and to dilute, or downblend, highly enriched uranium from weapons stockpiles which is now being redirected to become reactor fuel.

Back End

The back end of the nuclear fuel cycle, mostly spent fuel rods, contains fission products that emit beta and gamma radiation, and actinides that emit alpha particles, such as uranium-234, neptunium-237, plutonium-238 and americium-241, and even sometimes some neutron emitters such as californium (Cf). These isotopes are formed in nuclear reactors.

It is important to distinguish the processing of uranium to make fuel from the reprocessing of used fuel. Used fuel contains the highly radioactive products of fission.

Many of these are neutron absorbers, called neutron poisons in this context. These eventually build up to a level where they absorb so many neutrons that the chain reaction stops, even with the control rods completely removed. At that point the fuel has to be replaced in the reactor with fresh fuel, even though there is still a substantial quantity of uranium-235 and plutonium present. In the United States, this used fuel is usually "stored", while in other countries such as Russia, the United Kingdom, France, Japan and India, the fuel is reprocessed to remove the fission products, and the fuel can then be re-used, thus cutting costs, reducing health risks, saving time, and in general being far safer. The fission products removed from the fuel are a concentrated form of high-level waste as are the chemicals used in the process. While these countries reprocess the fuel carrying out single plutonium cycles, India is the only country known to be planning multiple plutonium recycling schemes.

Fuel Composition and Long Term Radioactivity

Long-lived radioactive waste from the back end of the fuel cycle is especially relevant when designing a complete waste management plan for spent nuclear fuel (SNF). When looking at long term radioactive decay, the actinides in the SNF have a significant influence due to their characteristically long half-lives. Depending on what a nuclear reactor is fueled with, the actinide composition in the SNF will be different.

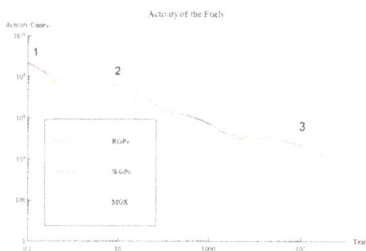

Total activity for three fuel types   Activity of U-233 for three fuel types

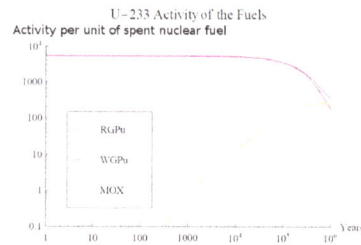

An example of this effect is the use of nuclear fuels with thorium. Th-232 is a fertile material that can undergo a neutron capture reaction and two beta minus decays, resulting in the production of fissile U-233. The SNF of a cycle with thorium will contain U-233. Its radioactive decay will strongly influence the long-term activity curve of the SNF around 1 million years. A comparison of the activity associated to U-233 for three different SNF types can be seen in the figure on the top right.

The burnt fuels are thorium with reactor-grade plutonium (RGPu), thorium with weapons-grade plutonium (WGPu) and Mixed Oxide fuel (MOX). For RGPu and WGPu, the initial amount of U-233 and its decay around 1 million years can be seen. This has an effect in the total activity curve of the three fuel types. The absence of U-233 and its daughter products in the MOX fuel results in a lower activity in region 3 of the figure on the bottom right, whereas for RGPu and WGPu the curve is maintained higher due to the presence of U-233 that has not fully decayed.

The use of different fuels in nuclear reactors results in different SNF composition, with varying activity curves.

Proliferation Concerns

Since uranium and plutonium are nuclear weapons materials, there have been proliferation concerns. Ordinarily (in spent nuclear fuel), plutonium is reactor-grade plutonium. In addition to plutonium-239, which is highly suitable for building nuclear weapons, it contains large amounts of undesirable contaminants: plutonium-240, plutonium-241, and plutonium-238. These isotopes are extremely difficult to separate, and more cost-effective ways of obtaining fissile material exist (e.g. uranium enrichment or dedicated plutonium production reactors).

High-level waste is full of highly radioactive fission products, most of which are relatively short-lived. This is a concern since if the waste is stored, perhaps in deep geological storage, over many years the fission products decay, decreasing the radioactivity of the waste and making the plutonium easier to access. The undesirable contaminant Pu-240 decays faster than the Pu-239, and thus the quality of the bomb material increases with time (although its quantity decreases during that time as well). Thus, some have argued, as time passes, these deep storage areas have the potential to become "plutonium mines", from which material for nuclear weapons can be acquired with relatively little difficulty. Critics of the latter idea have pointed out the difficulty of recovering useful material from sealed deep storage areas makes other methods preferable. Specifically, the high radioactivity and heat (80 C in surrounding rock) greatly increases the difficulty of mining a storage area, and the enrichment methods required have high capital costs.

Pu-239 decays to U-235 which is suitable for weapons and which has a very long half-life (roughly 10^9 years). Thus plutonium may decay and leave uranium-235. However, modern reactors are only moderately enriched with U-235 relative to U-238, so the U-238 continues to serve as a denaturation agent for any U-235 produced by plutonium decay.

One solution to this problem is to recycle the plutonium and use it as a fuel e.g. in fast reactors. In pyrometallurgical fast reactors, the separated plutonium and uranium are contaminated by actinides and cannot be used for nuclear weapons.

Nuclear Weapons Decommissioning

Waste from nuclear weapons decommissioning is unlikely to contain much beta or gamma activity other than tritium and americium. It is more likely to contain alpha-emitting actinides such as Pu-239 which is a fissile material used in bombs, plus some material with much higher specific activities, such as Pu-238 or Po.

In the past the neutron trigger for an atomic bomb tended to be beryllium and a high activity alpha emitter such as polonium; an alternative to polonium is Pu-238. For reasons of national security, details of the design of modern bombs are normally not released to the open literature.

Some designs might contain a radioisotope thermoelectric generator using Pu-238 to provide a long lasting source of electrical power for the electronics in the device.

It is likely that the fissile material of an old bomb which is due for refitting will contain decay products of the plutonium isotopes used in it, these are likely to include U-236 from Pu-240 impurities, plus some U-235 from decay of the Pu-239; due to the relatively long half-life of these Pu isotopes, these wastes from radioactive decay of bomb core material would be very small, and in any case, far less dangerous (even in terms of simple radioactivity) than the Pu-239 itself.

The beta decay of Pu-241 forms Am-241; the in-growth of americium is likely to be a greater problem than the decay of Pu-239 and Pu-240 as the americium is a gamma emitter (increasing external-exposure to workers) and is an alpha emitter which can cause the generation of heat. The plutonium could be separated from the americium by several different processes; these would include pyrochemical processes and aqueous/organic solvent extraction. A truncated PUREX type extraction process would be one possible method of making the separation. Naturally occurring uranium is not fissile because it contains 99.3% of U-238 and only 0.7% of U-235.

Legacy Waste

Due to historic activities typically related to radium industry, uranium mining, and military programs, there are numerous sites that contain or are contaminated with radioactivity. In the United States alone, the Department of Energy states there are "millions of gallons of radioactive waste" as well as "thousands of tons of spent nuclear fuel and material" and also "huge quantities of contaminated soil and water." Despite copious quantities of waste, the DOE has stated a goal of cleaning all presently contaminated sites successfully by 2025. The Fernald, Ohio site for example had "31 million pounds of uranium product", "2.5 billion pounds of waste", "2.75 million cubic yards of contaminated soil and debris", and a "223 acre portion of the underlying Great Miami Aquifer had uranium levels above drinking standards." The United States has at least 108 sites designated as areas that are contaminated and unusable, sometimes many thousands of acres. DOE wishes to clean or mitigate many or all by 2025, using the recently developed method of geomelting, however the task can be difficult and it acknowledges that some may never be completely remediated. In just one of these 108 larger designations, Oak Ridge National Laboratory, there were for example at least "167 known contaminant release sites" in one of the three subdivisions of the 37,000-acre (150 km^2) site. Some of the U.S. sites were smaller in nature, however, cleanup issues were simpler to address, and DOE has successfully completed cleanup, or at least closure, of several sites.

Medical

Radioactive medical waste tends to contain beta particle and gamma ray emitters. It can be divided into two main classes. In diagnostic nuclear medicine a number of short-lived gamma emitters such as technetium-99m are used. Many of these can be disposed of by leaving it to decay for a short time before disposal as normal waste. Other isotopes used in medicine, with half-lives in parentheses, include:

- Y-90, used for treating lymphoma (2.7 days)

- I-131, used for thyroid function tests and for treating thyroid cancer (8.0 days)

- Sr-89, used for treating bone cancer, intravenous injection (52 days)

- Ir-192, used for brachytherapy (74 days)

- Co-60, used for brachytherapy and external radiotherapy (5.3 years)

- Cs-137, used for brachytherapy, external radiotherapy (30 years)

Industrial

Industrial source waste can contain alpha, beta, neutron or gamma emitters. Gamma emitters are used in radiography while neutron emitting sources are used in a range of applications, such as oil well logging.

Naturally Occurring Radioactive Material (NORM)

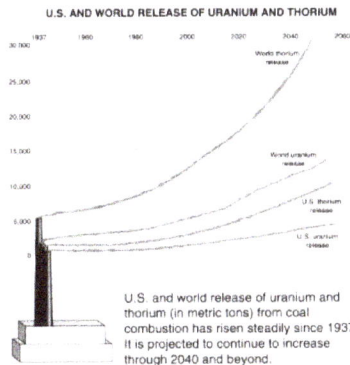

U.S. AND WORLD RELEASE OF URANIUM AND THORIUM

U.S. and world release of uranium and thorium (in metric tons) from coal combustion has risen steadily since 1937. It is projected to continue to increase through 2040 and beyond.

Annual release of uranium and thorium radioisotopes from coal combustion, predicted by ORNL to cumulatively amount to 2.9 million tons over the 1937–2040 period, from the combustion of an estimated 637 billion tons of coal worldwide.

Substances containing natural radioactivity are known as NORM. After human processing that exposes or concentrates this natural radioactivity (such as mining bringing coal to the surface or burning it to produce concentrated ash), it becomes technologically enhanced naturally occurring radioactive material (TENORM). A lot of this waste is alpha particle-emitting matter from the decay chains of uranium and thorium. The main source of radiation in the human body is potassium-40 (^{40}K), typically 17 milligrams in the body at a time and 0.4 milligrams/day intake. Most rocks, due to their components, have a low level of radioactivity. Usually ranging from 1 millisievert (mSv) to 13 mSv annually depending on location, average radiation exposure from natural radioisotopes is 2.0 mSv per person a year worldwide. This makes up the majority of typical total dosage (with mean annual exposure from other sources amounting to 0.6 mSv from medical tests averaged over the whole populace, 0.4 mSv from cosmic rays, 0.005 mSv from the legacy of past atmospheric nuclear testing, 0.005 mSv occupational exposure, 0.002 mSv from the Chernobyl disaster, and 0.0002 mSv from the nuclear fuel cycle).

TENORM is not regulated as restrictively as nuclear reactor waste, though there are no significant differences in the radiological risks of these materials.

Coal

Coal contains a small amount of radioactive uranium, barium, thorium and potassium, but, in the case of pure coal, this is significantly less than the average concentration of those elements in the Earth's crust. The surrounding strata, if shale or mudstone, often contain slightly more than average and this may also be reflected in the ash content of 'dirty' coals. The more active ash minerals become concentrated in the fly ash precisely because they do not burn well. The radioactivity of fly ash is about the same as black shale and is less than phosphate rocks, but is more of a concern because a small amount of the fly ash ends up in the atmosphere where it can be inhaled. According to U.S. NCRP reports, population exposure from 1000-MWe power plants amounts to 490 person-rem/year for coal power plants, 100 times as great as nuclear power plants (4.8 person-rem/year). (The exposure from the complete nuclear fuel cycle from mining to waste disposal is 136 person-rem/year; the corresponding value for coal use from mining to waste disposal is "probably unknown".)

Oil and Gas

Residues from the oil and gas industry often contain radium and its decay products. The sulfate scale from an oil well can be very radium rich, while the water, oil and gas from a well often contain radon. The radon decays to form solid radioisotopes which form coatings on the inside of pipework. In an oil processing plant the area of the plant where propane is processed is often one of the more contaminated areas of the plant as radon has a similar boiling point to propane.

Classification of Radioactive Waste

Classifications of radioactive waste varies by country. The IAEA, which publishes the Radioactive Waste Safety Standards (RADWASS), also plays a significant role.

Uranium Tailings

Uranium tailings are waste by-product materials left over from the rough processing of uranium-bearing ore. They are not significantly radioactive. Mill tailings are sometimes referred to as 11(e)2 wastes, from the section of the Atomic Energy Act of 1946 that defines them. Uranium mill tailings typically also contain chemically hazardous heavy metal such as lead and arsenic. Vast mounds of uranium mill tailings are left at many old mining sites, especially in Colorado, New Mexico, and Utah.

Removal of very low-level waste

Low-level Waste

Low level waste (LLW) is generated from hospitals and industry, as well as the nuclear fuel cycle. Low-level wastes include paper, rags, tools, clothing, filters, and other materials which contain small amounts of mostly short-lived radioactivity. Materials that originate from any region of an Active Area are commonly designated as LLW as a precautionary measure even if there is only a remote possibility of being contaminated with radioactive materials. Such LLW typically exhibits no higher radioactivity than one would expect from the same material disposed of in a non-active area, such as a normal office block.

Some high-activity LLW requires shielding during handling and transport but most LLW is suitable for shallow land burial. To reduce its volume, it is often compacted or incinerated before disposal. Low-level waste is divided into four classes: class A, class B, class C, and Greater Than Class C (GTCC).

Intermediate-level Waste

Intermediate-level waste (ILW) contains higher amounts of radioactivity and in general require shielding, but not cooling. Intermediate-level wastes includes resins, chemical sludge and metal nuclear fuel cladding, as well as contaminated materials from reactor decommissioning. It may be solidified in concrete or bitumen for disposal. As a general rule, short-lived waste (mainly non-fuel materials from reactors) is buried in shallow repositories, while long-lived waste (from fuel and fuel reprocessing) is deposited in geological repository. U.S. regulations do not define this category of waste; the term is used in Europe and elsewhere.

Spent fuel flasks are transported by railway in the United Kingdom. Each flask is constructed of 14 in (360 mm) thick solid steel and weighs in excess of 50 tons

High-level Waste

High-level waste (HLW) is produced by nuclear reactors. The exact definition of HLW differs internationally. After a nuclear fuel rod serves one fuel cycle and is removed from the core, it is considered HLW. Fuel rods contain fission products and transuranic elements generated in the reactor core. Spent fuel is highly radioactive and often hot. HLW accounts for over 95 percent of the total radioactivity produced in the process of nuclear electricity generation. The amount of HLW worldwide is currently increasing by about 12,000 metric tons every year, which is the equivalent to about 100 double-decker buses or a two-story structure with a footprint the size of a basketball court. A 1000-MW nuclear power plant produces about 27 tonnes of spent nuclear fuel (unreprocessed) every year. In 2010, there was very roughly estimated to be stored some 250,000 tons of nuclear HLW, that does not include amounts that have escaped into the environment from accidents or tests. Japan estimated to hold 17,000 tons of HLW in storage in 2015. HLW have been shipped to other countries to be stored or reprocessed, and in some cases, shipped back as active fuel.

The ongoing controversy over high-level radioactive waste disposal is a major constraint on the nuclear power's global expansion. Most scientists agree that the main proposed long-term solution is deep geological burial, either in a mine or a deep borehole. However, almost six decades after commercial nuclear energy began, no government has succeeded in opening such a repository for civilian high-level nuclear waste, although Finland is in the advanced stage of the construction of such facility, the Onkalo spent nuclear fuel repository. Reprocessing or recycling spent nuclear fuel options already available or under active development still generate waste and so are not a total solution, but can reduce the sheer quantity of waste, and there are many such active programs worldwide. Deep geological burial remains the only responsible way to deal with high-level nuclear waste. The Morris Operation is currently the only de facto high-level radioactive waste storage site in the United States.

Transuranic Waste

Transuranic waste (TRUW) as defined by U.S. regulations is, without regard to form or origin, waste that is contaminated with alpha-emitting transuranic radionuclides with half-lives greater than 20 years and concentrations greater than 100 nCi/g (3.7 MBq/kg), excluding high-level waste. Elements that have an atomic number greater than uranium are called transuranic ("beyond uranium"). Because of their long half-lives, TRUW is disposed more cautiously than either low- or intermediate-level waste. In the U.S., it arises mainly from weapons production, and consists of clothing, tools, rags, residues, debris and other items contaminated with small amounts of radioactive elements (mainly plutonium).

Under U.S. law, transuranic waste is further categorized into "contact-handled" (CH) and "remote-handled" (RH) on the basis of the radiation dose rate measured at the surface of the waste container. CH TRUW has a surface dose rate not greater than 200 mrem per hour (2 mSv/h), whereas RH TRUW has a surface dose rate of 200 mrem/h (2 mSv/h) or greater. CH TRUW does not have the very high radioactivity of high-level waste, nor its high heat generation, but RH TRUW can be highly radioactive, with surface dose rates up to 1,000,000 mrem/h (10,000 mSv/h). The U.S. currently disposes of TRUW generated from military facilities at the Waste Isolation Pilot Plant (WIPP) in a deep salt formation in New Mexico.

Prevention of Waste

A theoretical way to reduce waste accumulation is to phase out current reactors in favour of Generation IV Reactors or Liquid Fluoride Thorium Reactors, which output less waste per power generated. Fast reactors can theoretically consume some existing waste. The UK's Nuclear Decommissioning Authority published a position paper in 2014 on the progress on approaches to the management of separated plutonium, which summarises the conclusions of the work that NDA shared with UK government.

Management of Waste

Modern medium to high level transport container for nuclear waste

Of particular concern in nuclear waste management are two long-lived fission products, Tc-99 (half-life 220,000 years) and I-129 (half-life 15.7 million years), which dominate spent fuel radioactivity after a few thousand years. The most troublesome transuranic elements in spent fuel are Np-237 (half-life two million years) and Pu-239 (half-life 24,000 years). Nuclear waste requires sophisticated treatment and management to successfully isolate it from interacting with the biosphere. This usually necessitates treatment, followed by a long-term management strategy involving storage, disposal or transformation of the waste into a non-toxic form. Governments around the world are considering a range of waste management and disposal options, though there has been limited progress toward long-term waste management solutions.

In second half of 20th century, several methods of disposal of radioactive waste were investigated by nuclear nations, which are :

- "Long term above ground storage", not implemented.

- "Disposal in outer space" (for instance, inside the Sun), not implemented - as it would be currently too expensive.

- "Deep borehole disposal", not implemented.

- "Rock-melting", not implemented.

- "Disposal at subduction zones", not implemented.

- "Ocean disposal", done by the USSR, the United Kingdom, Switzerland, the United States, Belgium, France, The Netherlands, Japan, Sweden, Russia, Germany, Italy and South Korea. (1954–93) This is no longer permitted by international agreements.

- "Sub seabed disposal", not implemented, not permitted by international agreements.

- "Disposal in ice sheets", rejected in Antarctic Treaty

- "Direct injection", done by USSR and USA.

In the USA, waste management policy completely broke down with the ending of work on the incomplete Yucca Mountain Repository. At present there are 70 nuclear power plant sites where spent fuel is stored. A Blue Ribbon Commission was appointed by President Obama to look into future options for this and future waste. A Deep geological repository seems to be favored.

Initial Treatment of Waste

Vitrification

Long-term storage of radioactive waste requires the stabilization of the waste into a form which will neither react nor degrade for extended periods. It is theorized that one way to do this might be through vitrification. Currently at Sellafield the high-level waste (PUREX first cycle raffinate) is mixed with sugar and then calcined. Calcination involves passing the waste through a heated, rotating tube. The purposes of calcination are to evaporate the water from the waste, and de-nitrate the fission products to assist the stability of the glass produced.

The 'calcine' generated is fed continuously into an induction heated furnace with fragmented glass. The resulting glass is a new substance in which the waste products are bonded into the glass matrix when it solidifies. As a melt, this product is poured into stainless steel cylindrical containers ("cylinders") in a batch process. When cooled, the fluid solidifies ("vitrifies") into the glass. After being formed, the glass is highly resistant to water.

After filling a cylinder, a seal is welded onto the cylinder head. The cylinder is then washed. After being inspected for external contamination, the steel cylinder is stored, usually in an underground repository. In this form, the waste products are expected to be immobilized for thousands of years.

The glass inside a cylinder is usually a black glossy substance. All this work (in the United Kingdom) is done using hot cell systems. Sugar is added to control the ruthenium chemistry and to stop the formation of the volatile RuO_4 containing radioactive ruthenium isotopes. In the West, the glass is normally a borosilicate glass (similar to Pyrex), while in the former Soviet bloc it is normal to use a phosphate glass. The amount of fission products in the glass must be limited because some (palladium, the other Pt group metals, and tellurium) tend to form metallic phases which separate from the glass. Bulk vitrification uses electrodes to melt soil and wastes, which are then buried underground. In Germany a vitrification plant is in use; this is treating the waste from a small demonstration reprocessing plant which has since been closed down.

Ion Exchange

It is common for medium active wastes in the nuclear industry to be treated with ion exchange or other means to concentrate the radioactivity into a small volume. The much less radioactive bulk (after treatment) is often then discharged. For instance, it is possible to use a ferric hydroxide floc to remove radioactive metals from aqueous mixtures. After the radioisotopes are absorbed onto the ferric hydroxide, the resulting sludge can be placed in a metal drum before being mixed with cement to form a solid waste form. In order to get better long-term performance (mechanical stability) from such forms, they may be made from a mixture of fly ash, or blast furnace slag, and Portland cement, instead of normal concrete (made with Portland cement, gravel and sand).

Synroc

The Australian Synroc (synthetic rock) is a more sophisticated way to immobilize such waste, and this process may eventually come into commercial use for civil wastes (it is currently being developed for US military wastes). Synroc was invented by Prof Ted Ringwood (a geochemist) at the Australian National University. The Synroc contains pyrochlore and cryptomelane type minerals. The original form of Synroc (Synroc C) was designed for the liquid high level waste (PUREX raffinate) from a light water reactor. The main minerals in this Synroc are hollandite ($BaAl_2Ti_6O_{16}$), zirconolite ($CaZrTi_2O_7$) and perovskite ($CaTiO_3$). The zirconolite and perovskite are hosts for the actinides. The strontium and barium will be fixed in the perovskite. The caesium will be fixed in the hollandite.

Long Term Management of Waste

The time frame in question when dealing with radioactive waste ranges from 10,000 to 1,000,000 years, according to studies based on the effect of estimated radiation doses. Researchers suggest that forecasts of health detriment for such periods should be examined critically. Practical studies only consider up to 100 years as far as effective planning and cost evaluations are concerned. Long term behavior of radioactive wastes remains a subject for ongoing research projects in geoforecasting.

Above-ground Disposal

Dry cask storage typically involves taking waste from a spent fuel pool and sealing it (along with an inert gas) in a steel cylinder, which is placed in a concrete cylinder which acts as a radiation shield. It is a relatively inexpensive method which can be done at a central facility or adjacent to the source reactor. The waste can be easily retrieved for reprocessing.

Geologic Disposal

On Feb. 14, 2014, at the Waste Isolation Pilot Plant, radioactive materials leaked from a damaged storage drum (see photo). Analysis of several accidents, by DOE, have shown lack of a "safety culture" at the facility.

The process of selecting appropriate deep final repositories for high level waste and spent fuel is now under way in several countries with the first expected to be commissioned some time after 2010. The basic concept is to locate a large, stable geologic formation and use mining technology to excavate a tunnel, or large-bore tunnel boring machines (similar to those used to drill the Channel Tunnel from England to France) to drill a shaft 500 metres (1,600 ft) to 1,000 metres (3,300 ft) below the surface where rooms or vaults can be excavated for disposal of high-level radioactive waste. The goal is to permanently isolate nuclear waste from the human environment. Many people remain uncomfortable with the immediate stewardship cessation of this disposal system, suggesting perpetual management and monitoring would be more prudent.

Because some radioactive species have half-lives longer than one million years, even very low container leakage and radionuclide migration rates must be taken into account. Moreover, it may require more than one half-life until some nuclear materials lose enough radioactivity to cease being lethal to living things. A 1983 review of the Swedish radioactive waste disposal program by the National Academy of Sciences found that country's estimate of several hundred thousand years—perhaps up to one million years—being necessary for waste isolation "fully justified."

Ocean floor disposal of radioactive waste has been suggested by the finding that deep waters in the North Atlantic Ocean do not present an exchange with shallow waters for about 140 years based on oxygen content data recorded over a period of 25 years. They include burial beneath a stable abyssal plain, burial in a subduction zone that would slowly carry the waste downward into the Earth's mantle, and burial beneath a remote natural or human-made island. While these approaches all have merit and would facilitate an international solution to the problem of disposal of radioactive waste, they would require an amendment of the Law of the Sea.

Article 1 (Definitions), 7., of the 1996 Protocol to the Convention on the Prevention of Marine Pollution by Dumping of Wastes and Other Matter, (the London Dumping Convention) states:

> ""Sea" means all marine waters other than the internal waters of States, as well as the seabed and the subsoil thereof; it does not include sub-seabed repositories accessed only from land."

The proposed land-based subductive waste disposal method disposes of nuclear waste in a subduction zone accessed from land and therefore is not prohibited by international agreement. This method has been described as the most viable means of disposing of radioactive waste, and as the state-of-the-art as of 2001 in nuclear waste disposal technology. Another approach termed Remix & Return would blend high-level waste with uranium mine and mill tailings down to the level of the original radioactivity of the uranium ore, then replace it in inactive uranium mines. This approach has the merits of providing jobs for miners who would double as disposal staff, and of facilitating a cradle-to-grave cycle for radioactive materials, but would be inappropriate for spent reactor fuel in the absence of reprocessing, due to the presence in it of highly toxic radioactive elements such as plutonium.

Deep borehole disposal is the concept of disposing of high-level radioactive waste from nuclear reactors in extremely deep boreholes. Deep borehole disposal seeks to place the waste as much as 5 kilometres (3.1 mi) beneath the surface of the Earth and relies primarily on the immense natural geological barrier to confine the waste safely and permanently so that it should never pose a threat to the environment. The Earth's crust contains 120 trillion tons of thorium and 40 trillion tons of uranium (primarily at relatively trace concentrations of parts per million each adding up over the crust's $3 * 10^{19}$ ton mass), among other natural radioisotopes. Since the fraction of nuclides decaying per unit of time is inversely proportional to an isotope's half-life, the relative radioactivity of the lesser amount of human-produced radioisotopes (thousands of tons instead of trillions of tons) would diminish once the isotopes with far shorter half-lives than the bulk of natural radioisotopes decayed.

In January 2013, Cumbria county council rejected UK central government proposals to start work on an underground storage dump for nuclear waste near to the Lake District National Park. "For any host community, there will be a substantial community benefits package and worth hundreds of millions of pounds" said Ed Davey, Energy Secretary, but nonetheless, the local elected body voted 7–3 against research continuing, after hearing evidence from independent geologists that "the fractured strata of the county was impossible to entrust with such dangerous material and a hazard lasting millennia."

Transmutation

There have been proposals for reactors that consume nuclear waste and transmute it to other, less-harmful nuclear waste. In particular, the Integral Fast Reactor was a proposed nuclear reactor with a nuclear fuel cycle that produced no transuranic waste and in fact, could consume transuranic waste. It proceeded as far as large-scale tests, but was then canceled by the US Government. Another approach, considered safer but requiring more development, is to dedicate subcritical reactors to the transmutation of the left-over transuranic elements.

An isotope that is found in nuclear waste and that represents a concern in terms of proliferation is Pu-239. The estimated world total of plutonium in the year 2000 was of 1,645 metric tons, of which 210 metric tons had been separated by reprocessing. The large stock of plutonium is a result of its production inside uranium-fueled reactors and of the reprocessing of weapons-grade plutonium during the weapons program. An option for getting rid of this plutonium is to use it as a fuel in a traditional Light Water Reactor (LWR). Several fuel types with differing plutonium destruction efficiencies are under study.

Transmutation was banned in the US in April 1977 by President Carter due to the danger of plutonium proliferation, but President Reagan rescinded the ban in 1981. Due to the economic losses and risks, construction of reprocessing plants during this time did not resume. Due to high energy demand, work on the method has continued in the EU. This has resulted in a practical nuclear research reactor called Myrrha in which transmutation is possible. Additionally, a new research program called ACTINET has been started in the EU to make transmutation possible on a large, industrial scale. According to President Bush's Global Nuclear Energy Partnership (GNEP) of 2007, the US is now actively promoting research on transmutation technologies needed to markedly reduce the problem of nuclear waste treatment.

There have also been theoretical studies involving the use of fusion reactors as so called "actinide burners" where a fusion reactor plasma such as in a tokamak, could be "doped" with a small amount of the "minor" transuranic atoms which would be transmuted (meaning fissioned in the actinide case) to lighter elements upon their successive bombardment by the very high energy neutrons produced by the fusion of deuterium and tritium in the reactor. A study at MIT found that only 2 or 3 fusion reactors with parameters similar to that of the International Thermonuclear Experimental Reactor (ITER) could transmute the entire annual minor actinide production from all of the light water reactors presently operating in the United States fleet while simultaneously generating approximately 1 gigawatt of power from each reactor.

Re-use of Waste

Another option is to find applications for the isotopes in nuclear waste so as to re-use them. Already, caesium-137, strontium-90 and a few other isotopes are extracted for certain industrial applications such as food irradiation and radioisotope thermoelectric generators. While re-use does not eliminate the need to manage radioisotopes, it reduces the quantity of waste produced.

The Nuclear Assisted Hydrocarbon Production Method, Canadian patent application 2,659,302, is a method for the temporary or permanent storage of nuclear waste materials comprising the placing of waste materials into one or more repositories or boreholes constructed into an unconventional oil formation. The thermal flux of the waste materials fracture the formation and alters the chemical and/or physical properties of hydrocarbon material within the subterranean formation to allow removal of the altered material. A mixture of hydrocarbons, hydrogen, and/or other formation fluids is produced from the formation. The radioactivity of high-level radioactive waste affords proliferation resistance to plutonium placed in the periphery of the repository or the deepest portion of a borehole.

Breeder reactors can run on U-238 and transuranic elements, which comprise the majority of spent fuel radioactivity in the 1,000–100,000-year time span.

Space Disposal

Space disposal is attractive because it removes nuclear waste from the planet. It has significant disadvantages, such as the potential for catastrophic failure of a launch vehicle, which could spread radioactive material into the atmosphere and around the world. A high number of launches would be required because no individual rocket would be able to carry very much of the material relative to the total amount that needs to be disposed of. This makes the proposal impractical economically and it increases the risk of at least one or more launch failures. To further complicate matters, international agreements on the regulation of such a program would need to be established. Costs and inadequate reliability of modern rocket launch systems for space disposal has been one of the motives for interest in non-rocket space launch systems such as mass drivers, space elevators, and other proposals.

National Management Plans

Most countries are considerably ahead of the United States in developing plans for high-level radioactive waste disposal. Sweden and Finland are furthest along in committing to a particular disposal technology, while many others reprocess spent fuel or contract with France or Great Britain to do it, taking back the resulting plutonium and high-level waste. "An increasing backlog of plutonium from reprocessing is developing in many countries... It is doubtful that reprocessing makes economic sense in the present environment of cheap uranium."

In many European countries (e.g., Britain, Finland, the Netherlands, Sweden and Switzerland) the risk or dose limit for a member of the public exposed to radiation from a future high-level nuclear waste facility is considerably more stringent than that suggested by the International Commission on Radiation Protection or proposed in the United States. European limits are often more stringent than the standard suggested in 1990 by the International Commission on Radiation Protection by a factor of 20, and more stringent by a factor of ten than the standard proposed by the US Environmental Protection Agency (EPA) for Yucca Mountain nuclear waste repository for the first 10,000 years after closure.

The U.S. EPA's proposed standard for greater than 10,000 years is 250 times more permissive than the European limit. The U.S. EPA proposed a legal limit of a maximum of 3.5 millisieverts (350 millirem) each annually to local individuals after 10,000 years, which would be up to several percent of the exposure currently received by some populations in the highest natural background regions on Earth, though the U.S. DOE predicted that received dose would be much below that limit. Over a timeframe of thousands of years, after the most active short half-life radioisotopes decayed, burying U.S. nuclear waste would increase the radioactivity in the top 2000 feet of rock and soil in the United States (10 million km^2) by $\approx$ 1 part in 10 million over the cumulative amount of natural radioisotopes in such a volume, but the vicinity of the site would have a far higher concentration of artificial radioisotopes underground than such an average.

Mongolia

After serious opposition had arisen about plans and negotiations between Mongolia with Japan and the United States of America to build nuclear-waste facilities in Mongolia, Mongolia stopped all negotiations in September 2011. These negotiations had started after U.S. Deputy Secretary of Energy Daniel B. Poneman visited Mongolia in September, 2010. Talks took place in Washington DC between officials of Japan, the United States and Mongolia in February 2011. After this the United Arab Emirates (UAE), which wanted to buy nuclear fuel from Mongolia, joined in the negotiations. The talks were kept secret, and although The *Mainichi Daily News* reported on them in May, Mongolia officially denied the existence of these negotiations. However, alarmed by this news, Mongolian citizens protested against the plans, and demanded the government withdraw the plans and disclose information. The Mongolian President Tsakhiagiin Elbegdorj issued a presidential order on September 13 banning all negotiations with foreign governments or international organizations on nuclear-waste storage plans in Mongolia. The Mongolian government has accused the newspaper of distributing false claims around the world. After the presidential order, the Mongolian president fired the individual who was supposedly involved in these conversations.

Illegal Dumping

Authorities in Italy are investigating a 'Ndrangheta mafia clan accused of trafficking and illegally dumping nuclear waste. According to a whistleblower, a manager of the Italy's state energy research agency Enea paid the clan to get rid of 600 drums of toxic and radioactive waste from Italy, Switzerland, France, Germany, and the US, with Somalia as the destination, where the waste was buried after buying off local politicians. Former employees of Enea are suspected of paying the criminals to take waste off their hands in the 1980s and 1990s. Shipments to Somalia continued into the 1990s, while the 'Ndrangheta clan also blew up shiploads of waste, including radioactive hospital waste, and sending them to the sea bed off the Calabrian coast. According to the environmental group Legambiente, former members of the 'Ndrangheta have said that they were paid to sink ships with radioactive material for the last 20 years.

Accidents Involving Radioactive Waste

A few incidents have occurred when radioactive material was disposed of improperly, shielding during transport was defective, or when it was simply abandoned or even stolen from a waste store. In the Soviet Union, waste stored in Lake Karachay was blown over the area during a dust storm after the lake had partly dried out. At Maxey Flat, a low-level radioactive waste facility located in Kentucky, containment trenches covered with dirt, instead of steel or cement, collapsed under heavy rainfall into the trenches and filled with water. The water that invaded the trenches became radioactive and had to be disposed of at the Maxey Flat facility itself. In other cases of radioactive waste accidents, lakes or ponds with radioactive waste accidentally overflowed into the rivers during exceptional storms. In Italy, several radioactive waste deposits let material flow into river water, thus contaminating water for domestic use. In France, in the summer of 2008 numerous incidents happened; in one, at the Areva plant in Tricastin, it was reported that during a draining operation, liquid containing untreated uranium overflowed out of a faulty tank and about 75 kg of the radioactive material seeped into the ground and, from there, into two rivers nearby; in another case, over 100 staff were contaminated with low doses of radiation.

Scavenging of abandoned radioactive material has been the cause of several other cases of radiation exposure, mostly in developing nations, which may have less regulation of dangerous substances (and sometimes less general education about radioactivity and its hazards) and a market for scavenged goods and scrap metal. The scavengers and those who buy the material are almost always unaware that the material is radioactive and it is selected for its aesthetics or scrap value. Irresponsibility on the part of the radioactive material's owners, usually a hospital, university or military, and the absence of regulation concerning radioactive waste, or a lack of enforcement of such regulations, have been significant factors in radiation exposures. For an example of an accident involving radioactive scrap originating from a hospital see the Goiânia accident.

Transportation accidents involving spent nuclear fuel from power plants are unlikely to have serious consequences due to the strength of the spent nuclear fuel shipping casks.

On 15 December 2011 top government spokesman Osamu Fujimura of the Japanese government admitted that nuclear substances were found in the waste of Japanese nuclear facilities. Although Japan did commit itself in 1977 to these inspections in the safeguard agreement with the IAEA, the reports were kept secret for the inspectors of the International Atomic Energy Agency. Japan did start discussions with the IAEA about the large quantities of enriched uranium and plutonium that were discovered in nuclear waste cleared away by Japanese nuclear operators. At the press conference Fujimura said: "Based on investigations so far, most nuclear substances have been properly managed as waste, and from that perspective, there is no problem in safety management," But according to him, the matter was at that moment still being investigated.

Associated Hazard Warning Signs

The trefoil symbol used to indicate ionising radiation.

2007 ISO radioactivity danger symbol intended for IAEA Category 1, 2 and 3 sources defined as dangerous sources capable of death or serious injury.

The dangerous goods transport classification sign for radioactive materials

Deep Geological Repository

A deep geological repository is a nuclear waste repository excavated deep within a stable geologic environment (typically below 300 m or 1000 feet). It entails a combination of waste form, waste package, engineered seals and geology that is suited to provide a high level of long-term isolation and containment without future maintenance. The Waste Isolation Pilot Plant, under construction in the USA, is the only one which actually contains nuclear waste. But the facility suffered a chemical explosion which led to a serious radiological accident in 2014.

Technicians emplacing transuranic waste at the Waste Isolation Pilot Plant, near Carlsbad, New Mexico

Principles and Background

The most long-lived radioactive wastes, including spent nuclear fuel, must be contained and isolated from humans and the environment for a very long time. Disposal of these wastes in engineered facilities, or repositories, located deep underground in suitable geologic formations is seen as the reference solution. The International Panel on Fissile Materials has said:

It is widely accepted that spent nuclear fuel and high-level reprocessing and plutonium wastes require well-designed storage for periods ranging from tens of thousands to a million years, to minimize releases of the contained radioactivity into the environment. Safeguards are also required to ensure that neither plutonium nor highly enriched uranium is diverted to weapon use. There is general agreement that placing spent nuclear fuel in repositories hundreds of meters below the surface would be safer than indefinite storage of spent fuel on the surface.

However, even a storage space hundreds of metres below the ground would, in many parts of the developed world, have to be able to withstand the pressures of one or more future glaciations with thick sheets of ice resting on top of the rock, deforming it and creating internal strains, which is being taken into consideration by agencies preparing for long-term waste repositories in Sweden, Finland, Canada and some other countries that would have to expect a reneved ice age. '

Common elements of repositories include the radioactive waste, the containers enclosing the waste, other engineered barriers or seals around the containers, the tunnels housing the containers, and the geologic makeup of the surrounding area.

The ability of natural geologic barriers to isolate radioactive waste is demonstrated by the natural nuclear fission reactors at Oklo, Gabon. During their long reaction period about 5.4 tonnes of fission products as well as 1.5 tonnes of plutonium together with other transuranic elements were generated in the uranium ore body. This plutonium and the other transuranics remained immobile until the present day, a span of almost 2 billion years. This is quite remarkable in view of the fact that ground water had ready access to the deposits and they were not in a chemically inert form, such as glass.

Despite a long-standing agreement among many experts that geological disposal can be safe, technologically feasible and environmentally sound, a large part of the general public in many countries remains skeptical. One of the challenges facing the supporters of these efforts is to demonstrate confidently that a repository will contain wastes for so long that any releases that might take place in the future will pose no significant health or environmental risk.

Nuclear reprocessing does not eliminate the need for a repository, but reduces the volume, the long-term radiation hazard, and long-term heat dissipation capacity needed. Reprocessing does not eliminate the political and community challenges to repository siting.

Research

Deep geologic disposal has been studied for several decades, including laboratory tests, exploratory boreholes, and the construction and operation of underground research laboratories where large-scale in-situ tests are being conducted. Major underground test facilities are listed below.

Country	Facility name	Location	Geology	Depth	Status
Belgium	HADES Underground Research Facility	Mol	plastic clay	223 m	in operation 1982
Canada	AECL Underground Research Laboratory	Pinawa	granite	420 m	1990–2006
Finland	Onkalo	Olkiluoto	granite	400 m	under construction
France	Meuse/Haute Marne Underground Research Laboratory	Bure	mudstone	500 m	in operation 1999
Japan	Horonobe Underground Research Lab	Horonobe	sedimentary rock	500 m	under construction
Japan	Mizunami Underground Research Lab	Mizunami	granite	1000 m	under construction
Korea	Korea Underground Research Tunnel		granite	80 m	in operation 2006
Sweden	Äspö Hard Rock Laboratory	Oskarshamn	granite	450 m	in operation 1995
Switzerland	Grimsel Test Site	Grimsel Pass	granite	450 m	in operation 1984
Switzerland	Mont Terri Rock Laboratory	Mont Terri	claystone	300 m	in operation 1996
USA	Yucca Mountain nuclear waste repository	Nevada	tuff, ignimbrite	50 m	1997–2008

Repository Sites

A deep geological repository is a nuclear waste repository excavated deep within a stable geologic environment (typically below 300 m or 1000 feet). It entails a combination of waste form, waste package, engineered seals and geology that is suited to provide a high level of long-term isolation and containment without future maintenance. The Waste Isolation Pilot Plant, under construction in the USA, is the only one which actually contains nuclear waste. But the facility suffered a chemical explosion which led to a serious radiological accident in 2014.	Facility Name	Location	Waste	Geology	Depth	Status
Argentina	Sierra del Medio	Gastre		granite		under discussion
Belgium			high-level waste	plastic clay	~225 m	under discussion
Canada	OPG DGR	Ontario	200,000 m³ L&ILW	argillaceous limestone	680 m	license application 2011
Canada			spent fuel			under discussion
China						under discussion
Finland	VLJ	Olkiluoto	L&ILW	tonalite	60–100 m	in operation 1992
Finland		Loviisa	L&ILW	granite	120 m	in operation 1998
Finland	Onkalo	Olkiluoto	spent fuel	granite	400 m	under construction
France			high-level waste	mudstone	~500 m	siting
Germany	Schacht Asse II	Lower Saxony		salt dome	750 m	closed 1995

Germany	Morsleben	Saxony-An-halt	40,000 m³ L&ILW	salt dome	630 m	closed 1998
Germany	Gorleben	Lower Saxony	high-level waste	salt dome		proposed, on hold
Germany	Schacht Konrad	Lower Saxony	303,000 m³ L&ILW	sedimentary rock	800 m	under construction
Japan			high-level waste			under discussion
Korea	Gyeongju		L&ILW		80 m	under construction
Sweden	SFR	Forsmark	63,000 m³ L&ILW	granite	50 m	in operation 1988
Sweden		Forsmark	spent fuel	granite	450 m	license application 2011
Switzerland			high-level waste	clay		siting
United Kingdom			high-level waste			under discussion
USA	Waste Isolation Pilot Plant	New Mexico	transuranic waste	salt bed	655 m	in operation 1999
USA	Yucca Mountain Project	Nevada	70,000 ton HLW	ignimbrite	200–300 m	proposed, canceled 2010

The Current Situation at Certain Sites

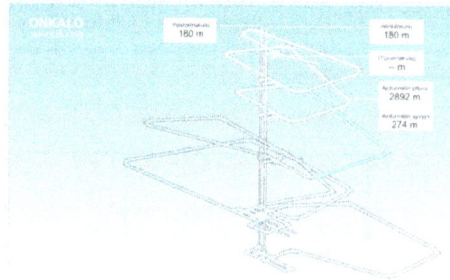

Schematic of a geologic repository under construction at Olkiluoto Nuclear Power Plant site, Finland

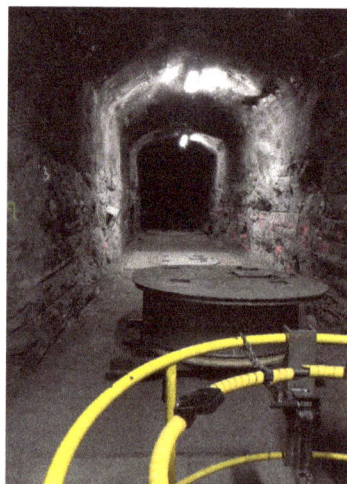

Demonstration tunnel in Olkiluoto.

The pit Asse II is a former salt mine in the mountain range of Asse in Lower Saxony/ Germany, that was allegedly used as a research mine since 1965. Between 1967 and

1978 radioactive waste was placed in storage. Research indicated that brine contaminated with radioactive caesium-137, plutonium and strontium was leaking from the mine since 1988 but was not reported until June 2008

The repository for radioactive waste Morsleben is a deep geological repository for radioactive waste in the rock salt mine Bartensleben in Morsleben, in Lower Saxony/ Germany that was used from 1972–1998. Since 2003 480,000 m³ (630,000 cu yd) of salt-concrete has been pumped into the pit to temporarily stabilize the upper levels. The salt dome is in the state of collapse.

The Waste Isolation Pilot Plant (WIPP) in the United States went into service in 1999 by putting the first cubic metres of transuranic radioactive waste in a deep layer of salt near Carlsbad, New Mexico.

There was a proposal for an international high level waste repository in Australia and Russia. However, since the proposal for a global repository in Australia (which has never produced nuclear power, and has one research reactor) was raised, domestic political objections have been loud and sustained, making such a facility in Australia unlikely.

In 1978 The U.S. Department of Energy began studying Yucca Mountain, within the secure boundaries of the Nevada Test Site in Nye County, Nevada, to determine whether it would be suitable for a long-term geologic repository for spent nuclear fuel and high-level radioactive waste. This project faced significant opposition and suffered delays due to litigation by The Agency for Nuclear Projects for the State of Nevada (Nuclear Waste Project Office) and others. The Obama Administration rejected use of the site in the 2009 United States Federal Budget proposal, which eliminated all funding except that needed to answer inquiries from the Nuclear Regulatory Commission, "while the Administration devises a new strategy toward nuclear waste disposal." On March 5, 2009, Energy Secretary Steven Chu told a Senate hearing the Yucca Mountain site is no longer viewed as an option for storing reactor waste.

Swedish KBS-3 capsule for nuclear waste.

In Germany, there is a political debate about the search for a final repository for radioactive waste, accompanied by loud protests, especially in the Gorleben village in the Wendland area, which was seen ideal for the final repository until 1990 because of its location in a remote, economically depressed corner of West Germany, next to the closed border to the former East Germany. After reunification, the village is now close to the center of the country, and is currently used for temporary storage of nuclear waste.

The process of selecting appropriate deep final repositories is now under way in several countries with the first expected to be commissioned some time after 2010. The Onkalo site in Finland is the furthest along the road to becoming operational, with waste burial currently scheduled to begin in 2020 (final approval is still missing). Sweden is also well advanced with plans for direct disposal of spent fuel, as its Parliament has decided that this is acceptably safe, using the KBS-3 technology.

The UK has been following the current path towards geological disposal since the 2008 Defra White Paper, entitled Managing Radioactive Waste Safely(MRWS). Unlike other developed countries the UK has placed the principle of voluntarism ahead of geological suitability. When seeking local council volunteers for stage 1 of the MRWS process only Allerdale and Copeland, within the county of Cumbria were volunteered by their councils. The same area that was previously examined and rejected in the 1990s. Stage 2 which was an initial unsuitability screening process was carried out by British Geological Survey (BGS) in 2010. This ruled out approximately 25% of the land area based on the presence of certain minerals and aquifers. There remains some controversy about this stage following accusations that the criteria were changed between the draft and final versions of this report, bringing the Solway Plain back into consideration, however the criteria were clearly published in the 2008 Defra White Paper, entitled Managing Radioactive Waste Safely (MRWS) 2 years prior to being applied.

In June 2012, the independent geologist advising the local West Cumbria MRWS Partnership group named three rock volumes that could be potentially suitable for geological disposal of nuclear waste. These are the Mercia Mudstone Group rocks between Silloth Abbeytown and Westnewton in North Cumbria, and the Ennerdale and Eskdale granites further south which lie within the Lake District National Park.

The decision on whether to proceed to the next stage is due in January 2013, and will be taken by just seven councillors, forming the Executive of Allerdale and another seven from Copeland. The ten member cabinet of Cumbria County Council have a veto which could prevent the search continuing.

In January 2013, Cumbria county council used its veto power and rejected UK central government proposals to start work on a production reactor nuclear waste repository near the Lake District National Park. "For any host community, there will be a substantial community benefits package and worth hundreds of millions of pounds" said Ed Davey, Energy Secretary, but nonetheless, the local elected administrative and governing body voted 7-3 against research continuing, after hearing evidence from independent geologists that "the fractured strata of the county was impossible to entrust with such dangerous material and a hazard lasting millennia."

Dry Cask Storage

Dry cask storage is a method of storing high-level radioactive waste, such as spent nuclear fuel that has already been cooled in the spent fuel pool for at least one year and often as much as ten years. Casks are typically steel cylinders that are either welded or bolted closed. The fuel rods inside are surrounded by inert gas. Ideally, the steel cylinder provides leak-tight containment of the spent fuel. Each cylinder is surrounded by additional steel, concrete, or other material to provide radiation shielding to workers and members of the public.

Dry cask storage area

There are various dry storage cask system designs. With some designs, the steel cylinders containing the fuel are placed vertically in a concrete vault; other designs orient the cylinders horizontally. The concrete vaults provide the radiation shielding. Other cask designs orient the steel cylinder vertically on a concrete pad at a dry cask storage site and use both metal and concrete outer cylinders for radiation shielding. Currently there is no long term permanent storage facility; dry cask storage is designed as an interim safer solution than spent fuel pool storage.

Some of the cask designs can be used for both storage and transportation. Three companies – Holtec International, NAC International and Areva-Transnuclear NUHOMS – are marketing Independent Spent Fuel Storage Installations (ISFSI's) based upon an unshielded multi-purpose canister which is transported and stored in on-site vertical or horizontal shielded storage modules constructed of steel and concrete.

Usage

During the 2000s, dry cask storage was used in the United States, Canada, Germany, Switzerland, Spain, Belgium, Sweden, the United Kingdom, Japan, Armenia, Argentina, Bulgaria, Czech Republic, Hungary, South Korea, Romania, Slovakia, Ukraine and Lithuania.

A similar system is also being implemented in Russia. However, it is based on 'storage compartments' in a single structure, rather than individual casks.

United States

In the late 1970s and early 1980s, the need for alternative storage in the United States began to grow when pools at many nuclear reactors began to fill up with stored spent fuel. As there was not a national nuclear storage facility in operation at the time, utilities began looking at options for storing spent fuel. Dry cask storage was determined to be a practical option for storage of spent fuel and preferable to leaving large concentrations of spent fuel in cooling tanks. The first dry storage installation in the US was licensed by the Nuclear Regulatory Commission (NRC) in 1986 at the Surry Nuclear Power Plant in Virginia, at 37°09′47″N 76°41′10″W / 37.1630°N 76.6861°W / 37.1630; -76.6861 (Surry Power Station spent fuel storage). Spent fuel is currently stored in dry cask systems at a growing number of power plant sites, and at an interim facility located at the Idaho National Laboratory near Idaho Falls, Idaho. The Nuclear Regulatory Commission estimates that many of the nuclear power plants in the United States will be out of room in their spent fuel pools by 2015, most likely requiring the use of temporary storage of some kind. Yucca Mountain was expected to open in 2017. However, on March 5, 2009, Energy Secretary Steven Chu reiterated in a Senate hearing that the Yucca Mountain site was no longer considered an option for storing reactor waste.

The 2008 NRC guideline calls for fuels to have spent at least five years in a storage pool before being moved to dry casks. The industry norm is about 10 years. The NRC describes the dry casks used in the US as "designed to resist floods, tornadoes, projectiles, temperature extremes, and other unusual scenarios."

As of the end of 2009, 13,856 metric tons of commercial spent fuel – or about 22 percent – were stored in dry casks.

In the 1990s, the NRC had to "take repeated actions to address defective welds on dry casks that led to cracks and quality assurance problems; helium had leaked into some casks, increasing temperatures and causing accelerated fuel corrosion".

With the zeroing of the budget for Yucca Mountain nuclear waste repository in Nevada, more nuclear waste is being loaded into sealed metal casks filled with inert gas. Many of these casks will be stored in coastal or lakeside regions where a salt air environment exists, and the Massachusetts Institute of Technology is studying how such dry casks perform in salt environments. Some hope that the casks can be used for 100 years, but cracking related to corrosion could occur in 30 years or less.

Canada

In Canada, above-ground dry storage has been used. Ontario Power Generation is in the process of constructing a Dry Storage Cask storage facility on its Darlington site, which will be similar in many respects to existing facilities at Pickering Nuclear Generating Station and Bruce Nuclear Generating Station. NB Power's Point Lepreau Nuclear Generating Station and Hydro-Québec's Gentilly Nuclear Generating Station also both operate dry storage facilities.

Germany

A centralized storage facility using dry casks is located at Ahaus. As of 2011, it housed 311 casks; 305 from the Thorium High Temperature Reactor, 3 from the Neckar-westheim Nuclear Power Plant, and 3 from the Gundremmingen Nuclear Power Plant. The transport from Gundremmingen to the Ahaus site met with considerable public protest and the power plant operators and the government later agreed to locate such casks at the powerplants.

CASTOR (*c*ask for *s*torage and *t*ransport *o*f *r*adioactive material) is a trademarked brand of dry casks used to store spent nuclear fuel (a type of nuclear waste). CASTORs are manufactured by GNS, a German provider of nuclear services.

CONSTOR is a cask used for transport and long-term storage of spent fuel and high-level waste manufactured by Gesellschaft für Nuklear-Service. Its inner and outer layers are steel, enclosing a layer of concrete.

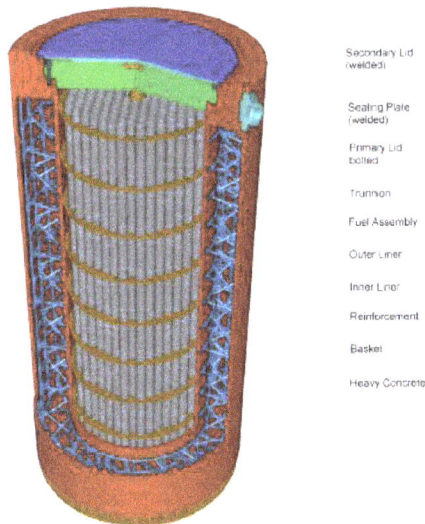

Multipurpose Constor Storage, Transport, and Disposal Cask

A 9-meter drop test of the V/TC model was conducted in 2004; the results conformed to expectations.

Bulgaria

In 2008, officials at the Kozloduy Nuclear Power Plant announced their intention to use 34 CONSTOR casks at the Kozloduy NPP site before the end of 2010.

Lithuania

Spent fuel from the now-closed Ignalina Nuclear Power Plant was placed in CASTOR and CONSTOR storage casks during the 2000s.

Russia

The Russian dry storage facility for spent nuclear fuel, the HOT-2 at Mining Chemical Combine in Zheleznogorsk, Krasnoyarsk Krai in Siberia, is not a 'cask' facility per se, as it is designed to accommodate the spent nuclear fuel (both VVER and RBMK) in a series of compartments. The structure of the facility is made up of monolithic reinforced concrete walls and top and bottom slabs, with the actual storage compartments formed by reinforced concrete partitions. The fuel is to be cooled by natural convection of air. The design capacity of the facility is 37,785 tonnes of uranium. It is now under construction and commissioning.

Ukraine

In Ukraine, a dry storage facility has been accepting spent fuel from the six-unit Zaporozhye Nuclear Power Plant (VVER-1000 reactors) since 2001, making it the longest-serving such facility in the former Soviet Union. The system was designed by the now-defunct Duke Engineering of the United States, with the storage casks being manufactured locally.

Another project is underway with Holtec International (again of the USA) to build a dry spent fuel storage facility at the 1986-accident-infamous Chernobyl Nuclear Power Plant (RBMK-1000 reactors). The project was initially started with Framatome (currently AREVA) of France, later suspended and terminated due to technical difficulties. Holtec was originally brought on board as a subcontractor to dehydrate the spent fuel, eventually taking over the entire project.

Nuclear Transmutation

Nuclear transmutation is the conversion of one chemical element or an isotope into another. Because any element (isotope) is defined by its number of protons (and neutrons) in its atoms, i.e. in the atomic nucleus, nuclear transmutation occurs in any process where this number is changed.

A transmutation can be achieved either by nuclear reactions (in which an outside particle reacts with a nucleus) or by radioactive decay (where no outside particle is needed).

Natural and Artificial Transmutation

Although all transmutation is caused by either decay or nuclear reaction, not all of these processes cause transmutation. E.g., gamma decay, internal conversion do not transmute the affected atom.

Natural transmutation by stellar nucleosynthesis in the past created most of the heavier chemical elements in the universe, see the corresponding section below.

One type of natural transmutation observable in the present occurs when certain radioactive elements present in nature spontaneously decay by a process that causes transmutation, such as alpha or beta decay. An example is the natural decay of potassium-40 to argon-40, which forms most of the argon in air. Also on Earth, natural transmutations from the different mechanism of *natural nuclear reactions* occur, due to cosmic ray bombardment of elements (for example, to form carbon-14), and also oc-casionally from natural neutron bombardment (for example, natural nuclear fission reactor).

Artificial transmutation may occur in machinery that has enough energy to cause changes in the nuclear structure of the elements. Such machines include particle accelerators and tokamak reactors. Conventional fission power reactors also cause artificial transmutation, not from the power of the machine, but by exposing elements to neutrons produced by fission from an artificially produced nuclear chain reaction. For instance, when a uranium atom is bombarded with slow neutrons, fission takes place. This releases, on average, 3 neutrons and a large amount of energy. The released neutrons then cause fission of other uranium atoms, until all of the available uranium is exhausted. This is called a chain reaction.

Artificial nuclear transmutation has been considered as a possible mechanism for reducing the volume and hazard of radioactive waste.

History

Alchemy

The term *transmutation* dates back to alchemy. Alchemists pursued the philosopher's stone, capable of chrysopoeia – the transformation of base metals into gold. While alchemists often understood chrysopoeia as a metaphor for a mystical, or religious process, some practitioners adopted a literal interpretation, and tried to make gold through physical experiment. The impossibility of the metallic transmutation had been debated amongst alchemists, philosophers and scientists since the Middle Ages. Pseudo-alchemical transmutation was outlawed and publicly mocked beginning in the fourteenth century. Alchemists like Michael Maier and Heinrich Khunrath wrote tracts exposing fraudulent claims of gold making. By the 1720s, there were no longer any respectable figures pursuing the physical transmutation of substances into gold. Antoine Lavoisier, in the 18th century, replaced the alchemical theory of elements with the modern theory of chemical elements, and John Dalton further developed the notion of atoms (from the alchemical theory of corpuscles) to explain various chemical processes. The disintegration of atoms is a distinct process involving much greater energies than could be achieved by alchemists.

Modern Physics

It was first consciously applied to modern physics by Frederick Soddy when he, along with Ernest Rutherford, discovered that radioactive thorium was converting itself into radium in 1901. At the moment of realization, Soddy later recalled, he shouted out: "Rutherford, this is transmutation!" Rutherford snapped back, "For Christ's sake, Soddy, don't call it *transmutation*. They'll have our heads off as alchemists."

Rutherford and Soddy were observing natural transmutation as a part of radioactive decay of the alpha decay type. However in 1919, Rutherford was able to accomplish transmutation of nitrogen into oxygen, using alpha particles directed at nitrogen $^{14}N + \alpha \rightarrow ^{17}O + p$. This was the first observation of a nuclear reaction, that is, a reaction in which particles from one decay are used to transform another atomic nucleus. Eventually, in 1932, a fully artificial nuclear reaction and nuclear transmutation was achieved by Rutherford's colleagues John Cockcroft and Ernest Walton, who used artificially accelerated protons against lithium-7 to split the nucleus into two alpha particles. The feat was popularly known as "splitting the atom," although it was not the modern nuclear fission reaction discovered in 1938 by Otto Hahn, Lise Meitner and their assistant Fritz Strassmann in heavy elements.

Later in the twentieth century the transmutation of elements within stars was elaborated, accounting for the relative abundance of heavier elements in the universe. Save for the first five elements, which were produced in the Big Bang and other cosmic ray processes, stellar nucleosynthesis accounted for the abundance of all elements heavier than boron. In their 1957 paper *Synthesis of the Elements in Stars*, William Alfred Fowler, Margaret Burbidge, Geoffrey Burbidge, and Fred Hoyle explained how the abundances of essentially all but the lightest chemical elements could be explained by the process of nucleosynthesis in stars.

It transpired that, under true nuclear transmutation, it is far easier to turn gold into lead than the reverse reaction, which was the one the alchemists had ardently pursued. Nuclear experiments have successfully transmuted lead into gold, but the expense far exceeds any gain. It would be easier to convert gold into lead via neutron capture and beta decay by leaving gold in a nuclear reactor for a long period of time.

Glenn Seaborg produced several thousand atoms of gold from bismuth, but at a net loss.

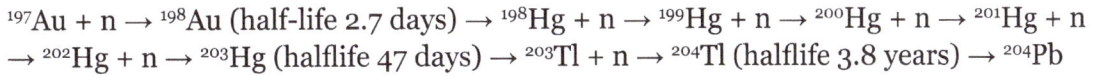

$^{197}Au + n \rightarrow {}^{198}Au$ (half-life 2.7 days) $\rightarrow {}^{198}Hg + n \rightarrow {}^{199}Hg + n \rightarrow {}^{200}Hg + n \rightarrow {}^{201}Hg + n \rightarrow {}^{202}Hg + n \rightarrow {}^{203}Hg$ (halflife 47 days) $\rightarrow {}^{203}Tl + n \rightarrow {}^{204}Tl$ (halflife 3.8 years) $\rightarrow {}^{204}Pb$

Transmutation in The Universe

The Big Bang is thought to be the origin of the hydrogen (including all deuterium) and helium in the universe. Hydrogen and helium together account for 98% of the mass of ordinary matter in the universe. The Big Bang also produced small amounts of lithium, beryllium and perhaps boron. More lithium, beryllium and boron were produced later, in a natural nuclear reaction, cosmic ray spallation.

Stellar nucleosynthesis is responsible for all of the other elements occurring naturally in the universe as stable isotopes and primordial nuclide, from carbon to plutonium. These occurred after the Big Bang, during star formation. Some lighter elements from carbon to iron were formed in stars and released into space by asymptotic giant branch (AGB) stars. These are a type of red giant that "puffs" off its outer atmosphere, containing some elements from carbon to nickel and iron. All elements with atomic weight greater than 64 atomic mass units are produced in supernova stars by means of nuclear reaction of lighter nuclei called neutron capture, which sub-divides into two processes: r-process and s-process.

The Solar System is thought to have condensed approximately 4.6 billion years before the present, from a cloud of hydrogen and helium containing heavier elements in dust grains formed previously by a large number of such stars. These grains contained the heavier elements formed by transmutation earlier in the history of the universe.

All of these natural processes of transmutation in stars are continuing today, in our own galaxy and in others. For example, the observed light curves of supernova stars such as SN 1987A show them blasting large amounts (comparable to the mass of Earth) of radioactive nickel and cobalt into space. However, little of this material reaches Earth. Most natural transmutation on the Earth today is mediated by cosmic rays (such as production of carbon-14) and by the radioactive decay of radioactive primordial nuclides left over from the initial formation of the solar system (such as potassium-40, uranium and thorium), plus the radioactive decay of products of these nucleides (radium, radon, polonium, etc.).

Artificial Transmutation of Nuclear Waste

Overview

Transmutation of transuranium elements (TRUs, i.e. actinides minus actinium to uranium) such as the isotopes of plutonium (about 1wt% in the Light Water Reactors' used nuclear fuel) or the minor actinides (MAs, i.e. neptunium, americium, and curium, about 0.1wt% each in LWRs' UNF) has the potential to help solve some problems posed by the management of radioactive waste by reducing the proportion of long-lived isotopes it contains. (This does not rule-out the need for a Deep Geological Repository (DGR) for High radioactive Level Waste (HLW).) When irradiated with fast neutrons in a nuclear reactor, these isotopes can undergo nuclear fission, destroying the original actinide isotope and producing a spectrum of radioactive and nonradioactive fission products.

Ceramic targets containing actinides can be bombarded with neutrons to induce transmutation reactions to remove the most difficult long-lived species. These can consist of actinide-containing solid solutions such as $(Am,Zr)N$, $(Am,Y)N$, $(Zr,Cm)O_2$, $(Zr,Cm,Am)O_2$, $(Zr,Am,Y)O_2$ or just actinide phases such as AmO_2, NpO_2, NpN, AmN mixed with some inert phases such as MgO, $MgAl_2O_4$, $(Zr,Y)O_2$, TiN and ZrN. The role of non-radioactive inert phases is mainly to provide stable mechanical behaviour to the target under neutron irradiation.

Reactor Types

For instance, plutonium can be reprocessed into MOX fuels and transmuted in standard reactors. The heavier elements could be transmuted in fast reactors, but probably more effectively in a subcritical reactor which is sometimes known as an energy amplifier and which was devised by Carlo Rubbia. Fusion neutron sources have also been proposed as well suited.

Fuel Types

There are several fuels that can incorporate plutonium in their initial composition at beginning of cycle (BOC) and have a smaller amount of this element at the end of cycle (EOC). During the cycle, plutonium can be burnt in a power reactor, generating electricity. This process is not only interesting from a power generation standpoint, but also due to its capability of consuming the surplus weapons grade plutonium from the weapons program and plutonium resulting of reprocessing UNF.

Mixed oxide fuel (MOX) is one of these. Its blend of oxides of plutonium and uranium constitutes an alternative to the low enriched uranium (LEU) fuel predominantly used in LWRs. Since uranium is present in MOX, although plutonium will be burnt, second generation plutonium will be produced through the radiative capture of U-238 and the two subsequent beta minus decays.

Fuels with plutonium and thorium are also an option. In these, the neutrons released in the fission of plutonium are captured by Th-232. After this radiative capture, Th-232 becomes Th-233, which undergoes two beta minus decays resulting in the production of the fissile isotope U-233. The radiative capture cross section for Th-232 is more than three times that of U-238, yielding a higher conversion to fissile fuel than that from U-238. Due to the absence of uranium in the fuel, there is no second generation plutonium produced, and the amount of plutonium burnt will be higher than in MOX fuels. However, U-233, which is fissile, will be present in the UNF. Weapons-grade and reactor-grade plutonium can be used in plutonium-thorium fuels, with weapons-grade plutonium being the one that shows a bigger reduction in the amount of Pu-239.

Reasoning Behind Transmutation

Isotopes of plutonium and other actinides tend to be long-lived with half-lives of many thousands of years, whereas radioactive fission products tend to be shorter-lived (most with half-lives of 30 years or less). From a waste management viewpoint, transmutation (or "burning" or "incineration") of actinides eliminates a very long-term radioactive hazard and replaces it with a much shorter-term one.

It is important to understand that the threat posed by a radioisotope is influenced by many factors including the physical (e.g. heat -infrared photon radiation-, which is an advantage for the storage or disposal af radioactive waste), chemical and biological properties of the element. For instance caesium has a relatively short biological half-life (1 to 4 months) while strontium and radium both have very long biological half-lives. As a result strontium-90 and radium are much more able to cause harm than caesium-137 when a given activity is ingested. Insert a brief calculation of doses

Many of the actinides are very radiotoxic because they have long biological half-lives and are alpha emitters. In transmutation the intention is to convert the actinides into fission products. The fission products are very radioactive, but the majority of the activity will decay away within a short time. The most worrying short-lived fission products are those that accumulate in the body, such as iodine-131 which accumulates in the thyroid gland, but it is hoped that by good design of the nuclear fuel and transmutation plant that such fission products can be isolated from humans and their environment and allowed to decay. In the medium term the fission products of highest concern are strontium-90 and caesium-137; both have a half-life of about 30 years. The caesium-137 is responsible for the majority of the external gamma dose experienced by workers in nuclear reprocessing plants and, in 2005, to workers at the Chernobyl site. When these medium-lived isotopes have decayed almost completely (usually after 10 half-lives) the remaining isotopes will pose a much smaller threat.

Long-lived Fission Products (LLFP)

Long-lived fission products				
Prop: **Unit:**	$t_{1/2}$ **(Ma)**	**Yield** **(%)**	**Q *** **(keV)**	**βγ** *****
^{99}Tc	0.211	6.1385	294	β

^{126}Sn	0.230	0.1084	4050	βγ
^{79}Se	0.327	0.0447	151	β
^{93}Zr	1.53	5.4575	91	βγ
^{135}Cs	2.3	6.9110	269	β
^{107}Pd	6.5	1.2499	33	β
^{129}I	15.7	0.8410	194	βγ
Hover underlined: more info				

Medium-lived fission products				
Prop: Unit:	$t_{1/2}$ (a)	Yield (%)	Q * (keV)	βγ *
^{155}Eu	4.76	0.0803	252	βγ
^{85}Kr	10.76	0.2180	687	βγ
^{113m}Cd	14.1	0.0008	316	β
^{90}Sr	28.9	4.505	2826	**β**
^{137}Cs	30.23	6.337	1176	**βγ**
^{121m}Sn	43.9	0.00005	390	βγ
^{151}Sm	96.6	0.5314	77	β

Some radioactive fission products can be converted into shorter-lived radioisotopes by transmutation. Transmutation of all fission products with half-life greater than one year is studied in Grenoble, with varying results.

Sr-90 and Cs-137, with half-lives of about 30 years, are the largest radiation (including heat) emitters in used nuclear fuel on a scale of decades to ~305 years (Sn-121m is insignificant because of the low yield), and are not easily transmuted because they have low neutron absorption cross sections. Instead, they should simply be stored until they decay. Given that this length of storage is necessary, the fission products with shorter half-lives can also be stored until they decay.

The next longer-lived fission product is Sm-151, which has a half-life of 90 years, and is such a good neutron absorber that most of it is transmuted while the nuclear fuel is still being used; however, effectively transmuting the remaining Sm-151 in nuclear waste would require separation from other isotopes of samarium. Given the smaller quantities and its low-energy radioactivity, Sm-151 is less dangerous than Sr-90 and Cs-137 and can also be left to decay for ~970 years.

Finally, there are 7 long-lived fission products. They have much longer half-lives in the range 211,000 years to 15.7 million years. Two of them, Tc-99 and I-129, are mobile enough in the environment to be potential dangers, are free or mostly free of mixture with stable isotopes of the same element, and have neutron cross sections that are small but adequate to support transmutation. Also, Tc-99 can substitute for U-238 in supplying Doppler broadening for negative feedback for reactor stability. Most studies of proposed transmutation schemes have assumed ^{99}Tc, ^{129}I, and TRUs as the targets for transmutation, with other fission products, activation products, and possibly reprocessed uranium remaining as waste.

Of the remaining 5 long-lived fission products, Se-79, Sn-126 and Pd-107 are produced only in small quantities (at least in today's thermal neutron, U-235-burning light water reactors) and the last two should be relatively inert. The other two, Zr-93 and Cs-135, are produced in larger quantities, but also not highly mobile in the environment. They are also mixed with larger quantities of other isotopes of the same element.

Nuclear Reprocessing

Nuclear reprocessing technology was developed to chemically separate and recover fissionable plutonium from irradiated nuclear fuel. Reprocessing serves multiple purposes, whose relative importance has changed over time. Originally, reprocessing was used solely to extract plutonium for producing nuclear weapons. With the commercialization of nuclear power, the reprocessed plutonium was recycled back into MOX nuclear fuel for thermal reactors. The reprocessed uranium, which constitutes the bulk of the spent fuel material, can in principle also be re-used as fuel, but that is only economical when uranium prices are high. Finally, a breeder reactor is not restricted to using recycled plutonium and uranium. It can employ all the actinides, closing the nuclear fuel cycle and potentially multiplying the energy extracted from natural uranium by about 60 times.

Nuclear reprocessing reduces the volume of high-level waste, but by itself does not reduce radioactivity or heat generation and therefore does not eliminate the need for a geological waste repository. Reprocessing has been politically controversial because of the potential to contribute to nuclear proliferation, the potential vulnerability to nuclear terrorism, the political challenges of repository siting (a problem that applies equally to direct disposal of spent fuel), and because of its high cost compared to the once-through fuel cycle. In the United States, the Obama administration stepped back from President Bush's plans for commercial-scale reprocessing and reverted to a program focused on reprocessing-related scientific research. Nuclear fuel reprocessing is performed routinely in Europe, Russia and Japan.

Separated Components and Disposition

The potentially useful components dealt with in nuclear reprocessing comprise specific actinides (plutonium, uranium, and some minor actinides). The lighter elements components include fission products, activation products, and cladding.

material	disposition
plutonium, minor actinides, reprocessed uranium	fission in fast, fusion, or subcritical reactor
reprocessed uranium, cladding, filters	less stringent storage as intermediate-level waste
long-lived fission and activation products	nuclear transmutation or geological repository
medium-lived fission products ^{137}Cs and ^{90}Sr	medium-term storage as high-level waste
useful radionuclides and noble metals	industrial and medical uses

History

The first large-scale nuclear reactors were built during World War II. These reactors were designed for the production of plutonium for use in nuclear weapons. The only reprocessing required, therefore, was the extraction of the plutonium (free of fission-product contamination) from the spent natural uranium fuel. In 1943, several methods were proposed for separating the relatively small quantity of plutonium from the uranium and fission products. The first method selected, a precipitation process called the bismuth phosphate process, was developed and tested at the Oak Ridge National Laboratory (ORNL) between 1943 and 1945 to produce quantities of plutonium for evaluation and use in the US weapons programs. ORNL produced the first macroscopic quantities (grams) of separated plutonium with these processes.

The bismuth phosphate process was first operated on a large scale at the Hanford Site, in the later part of 1944. It was successful for plutonium separation in the emergency situation existing then, but it had a significant weakness: the inability to recover uranium.

The first successful solvent extraction process for the recovery of pure uranium and plutonium was developed at ORNL in 1949. The PUREX process is the current method of extraction. Separation plants were also constructed at Savannah River Site and a smaller plant at West Valley Reprocessing Plant which closed by 1972 because of its inability to meet new regulatory requirements.

Reprocessing of civilian fuel has long been employed at the COGEMA La Hague site in France, the Sellafield site in the United Kingdom, the Mayak Chemical Combine in Russia, and at sites such as the Tokai plant in Japan, the Tarapur plant in India, and briefly at the West Valley Reprocessing Plant in the United States.

In October 1976, concern of nuclear weapons proliferation (especially after India demonstrated nuclear weapons capabilities using reprocessing technology) led President Gerald Ford to issue a Presidential directive to indefinitely suspend the commercial reprocessing and recycling of plutonium in the U.S. On 7 April 1977, President Jimmy Carter banned the reprocessing of commercial reactor spent nuclear fuel. The key issue driving this policy was the serious threat of nuclear weapons proliferation by diversion of plutonium from the civilian fuel cycle, and to encourage other nations to follow the USA lead. After that, only countries that already had large investments in reprocessing infrastructure continued to reprocess spent nuclear fuel. President Reagan lifted the ban in 1981, but did not provide the substantial subsidy that would have been necessary to start up commercial reprocessing.

In March 1999, the U.S. Department of Energy (DOE) reversed its policy and signed a contract with a consortium of Duke Energy, COGEMA, and Stone & Webster (DCS) to design and operate a mixed oxide (MOX) fuel fabrication facility. Site preparation at the Savannah River Site (South Carolina) began in October 2005. In 2011 the New York Times reported "...11 years after the government awarded a construction contract, the cost of the project has soared to nearly $5 billion. The vast concrete and steel structure is a half-finished hulk, and the government has yet to find a single customer, despite offers of lucrative subsidies." TVA (currently the most likely customer) said in April 2011 that it would delay a decision until it could see how MOX fuel performed in the nuclear accident at Fukushima Daiichi.

Separation Technologies

Water and Organic Solvents

PUREX

PUREX, the current standard method, is an acronym standing for *Plutonium and Uranium Recovery by EXtraction*. The PUREX process is a liquid-liquid extraction method used to reprocess spent nuclear fuel, to extract uranium and plutonium, independent of each other, from the fission products. This is the most developed and widely used process in the industry at present. When used on fuel from commercial power reactors the plutonium extracted typically contains too much Pu-240 to be considered "weapons-grade" plutonium, ideal for use in a nuclear weapon. Nevertheless, highly reliable nuclear weapons can be built at all levels of technical sophistication using reactor-grade plutonium. Moreover, reactors that are capable of refueling frequently can be used to produce weapon-grade plutonium, which can later be recovered using PUREX. Because of this, PUREX chemicals are monitored.

Plutonium Processing

Modifications of PUREX

UREX

The PUREX process can be modified to make a UREX (URanium EXtraction) process which could be used to save space inside high level nuclear waste disposal sites, such as the Yucca Mountain nuclear waste repository, by removing the uranium which makes up the vast majority of the mass and volume of used fuel and recycling it as reprocessed uranium.

The UREX process is a PUREX process which has been modified to prevent the plutonium from being extracted. This can be done by adding a plutonium reductant before the first metal extraction step. In the UREX process, ~99.9% of the uranium and >95% of technetium are separated from each other and the other fission products and actinides. The key is the addition of acetohydroxamic acid (AHA) to the extraction and scrub sections of the process. The addition of AHA greatly diminishes the extractability of plutonium and neptunium, providing somewhat greater proliferation resistance than with the plutonium extraction stage of the PUREX process.

TRUEX

Adding a second extraction agent, octyl(phenyl)-N, N-dibutyl carbamoylmethyl phosphine oxide(CMPO) in combination with tributylphosphate, (TBP), the PUREX process can be turned into the TRUEX (TRansUranic EXtraction) process. TRUEX was invented in the USA by Argonne National Laboratory and is designed to remove the transuranic metals (Am/Cm) from waste. The idea is that by lowering the alpha activity of the waste, the majority of the waste can then be disposed of with greater ease. In common with PUREX this process operates by a solvation mechanism.

DIAMEX

As an alternative to TRUEX, an extraction process using a malondiamide has been devised. The DIAMEX (DIAMideEXtraction) process has the advantage of avoiding the formation of organic waste which contains elements other than carbon, hydrogen, nitrogen, and oxygen. Such an organic waste can be burned without the formation of acidic gases which could contribute to acid rain (although the acidic gases could be recovered by a scrubber). The DIAMEX process is being worked on in Europe by the French CEA. The process is sufficiently mature that an industrial plant could be constructed with the existing knowledge of the process. In common with PUREX this process operates by a solvation mechanism.

SANEX

Selective ActiNide EXtraction. As part of the management of minor actinides it has been proposed that the lanthanides and trivalent minor actinides should be removed from the PUREX raffinate by a process such as DIAMEX or TRUEX. In order to allow the actinides such as americium to be either reused in industrial sources or used as fuel, the lanthanides must be removed. The lanthanides have large neutron cross sections and hence they would poison a neutron driven nuclear reaction. To date the extraction system for the SANEX process has not been defined, but currently several different research groups are working towards a process. For instance the French CEA is working on a bis-triazinyl pyridine (BTP) based process. Other systems such as the dithiophosphinic acids are being worked on by some other workers.

UNEX

The *UNiversal* EXtraction process was developed in Russia and the Czech Republic; it is designed to completely remove the most troublesome radioisotopes (Sr, Cs and minor actinides) from the raffinate remaining after the extraction of uranium and plutonium from used nuclear fuel. The chemistry is based upon the interaction of caesium and strontium with polyethylene glycol) and a cobalt carborane anion (known as chlorinated cobalt dicarbollide). The actinides are extracted by CMPO, and the diluent is a polar aromatic such as nitrobenzene. Other dilents such as *meta*-nitrobenzotrifluoride and phenyl trifluoromethyl sulfone have been suggested as well.

Electrochemical Methods

An exotic method using electrochemistry and ion exchange in ammonium carbonate has been reported.

Obsolete Methods

Bismuth Phosphate

The bismuth phosphate process is an obsolete process that adds significant unnecessary material to the final radioactive waste. The bismuth phosphate process has been replaced by solvent extraction processes. The bismuth phosphate process was designed to extract plutonium from aluminium-clad nuclear fuel rods, containing uranium. The fuel was decladded by boiling it in caustic soda. After decladding, the uranium metal was dissolved in nitric acid.

The plutonium at this point is in the +4 oxidation state. It was then precipitated out of the solution by the addition of bismuth nitrate and phosphoric acid to form the bismuth phosphate. The plutonium was coprecipitated with this. The supernatant liquid (containing many of the fission products) was separated from the solid. The precipitate was then dissolved in nitric acid before the addition of an oxidant such as potassium permanganate which converted the plutonium to PuO_2^{2+} (Pu VI), then a dichromate salt was added to maintain the plutonium in the +6 oxidation state.

The bismuth phosphate was next re-precipitated leaving the plutonium in solution. Then an iron (II) salt such as ferrous sulfate was added, and the plutonium re-precipitated again using a bismuth phosphate carrier precipitate. Then lanthanum salts and fluoride were added to create solid lanthanum fluoride which acted as a carrier for the plutonium. This was converted to the oxide by the action of an alkali. The lanthanum plutonium oxide was next collected and extracted with nitric acid to form plutonium nitrate.

Hexone or Redox

This is a liquid-liquid extraction process which uses methyl isobutyl ketone as the extractant. The extraction is by a *solvation* mechanism. This process has the disadvantage of requiring the use of a salting-out reagent (aluminium nitrate) to increase the nitrate concentration in the aqueous phase to obtain a reasonable distribution ratio (D value). Also, hexone is degraded by concentrated nitric acid. This process has been replaced by the PUREX process.

$$Pu^{4+} + 4\ NO_3^- + 2S \rightarrow [Pu(NO_3)_4 S_2]$$

Butex, β,β'-dibutyoxydiethyl Ether

A process based on a solvation extraction process using the triether extractant named above. This process has the disadvantage of requiring the use of a salting-out reagent (aluminium nitrate) to increase the nitrate concentration in the aqueous phase to obtain a reasonable distribution ratio. This process was used at Windscale many years ago. This process has been replaced by PUREX.

Pyroprocessing

Pyroprocessing is a generic term for high-temperature methods. Solvents are molten salts (e.g. LiCl+KCl or LiF+CaF2) and molten metals (e.g. cadmium, bismuth, magnesium) rather than water and organic compounds. Electrorefining, distillation, and solvent-solvent extraction are common steps.

These processes are not currently in significant use worldwide, but they have been researched and developed at Argonne National Laboratory and elsewhere.

Advantages

- The principles behind them are well understood, and no significant technical barriers exist to their adoption.

- Readily applied to high-burnup spent fuel and requires little cooling time, since the operating temperatures are high already.

- Does not use solvents containing hydrogen and carbon, which are neutron moderators creating risk of criticality accidents and can absorb the fission product tritium and the activation product carbon-14 in dilute solutions that cannot be separated later.

- Alternatively, voloxidation can remove 99% of the tritium from used fuel and recover it in the form of a strong solution suitable for use as a supply of tritium.

- More compact than aqueous methods, allowing on-site reprocessing at the reactor site, which avoids transportation of spent fuel and its security issues, instead storing a much smaller volume of fission products on site as high-level waste until decommissioning. For example, the Integral Fast Reactor and Molten Salt Reactor fuel cycles are based on on-site pyroprocessing.

- It can separate many or even all actinides at once and produce highly radioactive fuel which is harder to manipulate for theft or making nuclear weapons. (However, the difficulty has been questioned.) In contrast the PUREX process was designed to separate plutonium only for weapons, and it also leaves the minor actinides (americium and curium) behind, producing waste with more long-lived radioactivity.

- Most of the radioactivity in roughly 10^2 to 10^5 years after the use of the nuclear fuel is produced by the actinides, since there are no fission products with half-lives in this range. These actinides can fuel fast reactors, so extracting and reusing (fissioning) them reduces the long-term radioactivity of the wastes.

Disadvantages

- Reprocessing as a whole is not currently (2005) in favor, and places that do reprocess already have PUREX plants constructed. Consequently, there is little demand for new pyrometalurgical systems, although there could be if the Generation IV reactor programs become reality.

- The used salt from pyroprocessing is less suitable for conversion into glass than the waste materials produced by the PUREX process.

- If the goal is to reduce the longevity of spent nuclear fuel in burner reactors, then better recovery rates of the minor actinides need to be achieved.

Electrolysis

PYRO-A and -B for IFR

These processes were developed by Argonne National Laboratory and used in the Integral Fast Reactor project.

PYRO-A is a means of separating actinides (elements within the actinide family, generally heavier than U-235) from non-actinides. The spent fuel is placed in an anode basket which is immersed in a molten salt electrolyte. An electric current is applied, causing the uranium metal (or sometimes oxide, depending on the spent fuel) to plate out on a solid metal cathode while the other actinides (and the rare earths) can be absorbed into a liquid cadmium cathode. Many of the fission products (such as caesium, zirconium and strontium) remain in the salt. As alternatives to the molten cadmium electrode it is possible to use a molten bismuth cathode, or a solid aluminium cathode.

As an alternative to electrowinning, the wanted metal can be isolated by using a molten alloy of an electropositive metal and a less reactive metal.

Since the majority of the long term radioactivity, and volume, of spent fuel comes from actinides, removing the actinides produces waste that is more compact, and not nearly as dangerous over the long term. The radioactivity of this waste will then drop to the level of various naturally occurring minerals and ores within a few hundred, rather than thousands of, years.

The mixed actinides produced by pyrometallic processing can be used again as nuclear fuel, as they are virtually all either fissile, or fertile, though many of these materials would require a fast breeder reactor in order to be burned efficiently. In a thermal neutron spectrum, the concentrations of several heavy actinides (curium-242 and plutonium-240) can become quite high, creating fuel that is substantially different from the usual uranium or mixed uranium-plutonium oxides (MOX) that most current reactors were designed to use.

Another pyrochemical process, the PYRO-B process, has been developed for the processing and recycling of fuel from a transmuter reactor (a fast breeder reactor designed to convert transuranic nuclear waste into fission products). A typical transmuter fuel is free from uranium and contains recovered transuranics in an inert matrix such as metallic zirconium. In the PYRO-B processing of such fuel, an electrorefining step is used to separate the residual transuranic elements from the fission products and recycle the transuranics to the reactor for fissioning. Newly generated technetium and iodine are extracted for incorporation into transmutation targets, and the other fission products are sent to waste.

Voloxidation

Voloxidation (for *volumetric oxidation*) involves heating oxide fuel with oxygen, sometimes with alternating oxidation and reduction, or alternating oxidation by ozone to uranium trioxide with decomposition by heating back to triuranium octoxide. A major purpose is to capture tritium as tritiated water vapor before further processing where it would be difficult to retain the tritium. Other volatile elements leave the fuel and must be recovered, especially iodine, technetium, and carbon-14. Voloxidation also breaks up the fuel or increases its surface area to enhance penetration of reagents in following reprocessing steps.

Volatilization in Isolation

Simply heating spent oxide fuel in an inert atmosphere or vacuum at a temperature between 700 °C and 1000 °C as a first reprocessing step can remove several volatile elements, including caesium whose isotope caesium-137 emits about half of the heat produced by the spent fuel over the following 100 years of cooling (however, most of the other half is from strontium-90 which remains). The estimated overall mass balance for 20,000 grams of processed fuel with 2,000 grams of cladding is:

	Input	Residue	Zeolite filter	Carbon filter	Particle filters
Palladium	28	14	14		
Tellurium	10	5	5		
Molybdenum	70		70		
Caesium	46		46		
Rubidium	8		8		
Silver	2		2		
Iodine	4			4	
Cladding	2000	2000			
Uranium	19218	19218			?
Others	614	614			?
Total	22000	21851	145	4	0

Tritium is not mentioned in this paper.

Fluoride Volatility

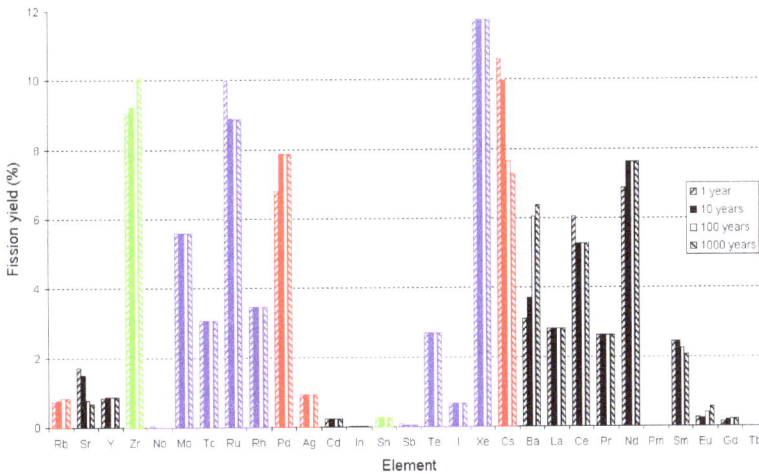

Blue elements have volatile fluorides or are already volatile; green elements do not but have volatile chlorides; red elements have neither, but the elements themselves or their oxides are volatile at very high temperatures. Yields at $10^{0,1,2,3}$ years after fission, not considering later neutron capture, fraction of 100% not 200%. Beta decay Kr-85→Rb, Sr-90→Zr, Ru-106→Pd, Sb-125→Te, Cs-137→Ba, Ce-144→Nd, Sm-151→Eu, Eu-155→Gd visible.

In the fluoride volatility process, fluorine is reacted with the fuel. Fluorine is so much more reactive than even oxygen that small particles of ground oxide fuel will burst into flame when dropped into a chamber full of fluorine. This is known as flame fluorination; the heat produced helps the reaction proceed. Most of the uranium, which makes up the bulk of the fuel, is converted to uranium hexafluoride, the form of uranium used in uranium enrichment, which has a very low boiling point. Technetium, the main long-lived fission product, is also efficiently converted to its volatile hexafluoride. A few other elements also form similarly volatile hexafluorides, pentafluorides, or heptafluorides. The volatile fluorides can be separated from excess fluorine by condensation, then separated from each other by fractional distillation or selective reduction. Uranium hexafluoride and technetium hexafluoride have very similar boiling points and vapor pressures, which makes complete separation more difficult.

Many of the fission products volatilized are the same ones volatilized in non-fluorinated, higher-temperature volatilization, such as iodine, tellurium and molybdenum; notable differences are that technetium is volatilized, but caesium is not.

Some transuranium elements such as plutonium, neptunium and americium can form volatile fluorides, but these compounds are not stable when the fluorine partial pressure is decreased. Most of the plutonium and some of the uranium will initially remain in ash which drops to the bottom of the flame fluorinator. The plutonium-uranium ratio in the ash may even approximate the composition needed for fast neutron reactor fuel. Further fluorination of the ash can remove all the uranium, neptunium, and plutonium as volatile fluorides; however, some other minor actinides may not form volatile fluorides and instead remain with the alkaline fission products. Some noble metals may not form fluorides at all, but remain in metallic form; however ruthenium hexafluoride is relatively stable and volatile.

Distillation of the residue at higher temperatures can separate lower-boiling transition metal fluorides and alkali metal (Cs, Rb) fluorides from higher-boiling lanthanide and alkaline earth metal (Sr, Ba) and yttrium fluorides. The temperatures involved are much higher, but can be lowered somewhat by distilling in a vacuum. If a carrier salt like lithium fluoride or sodium fluoride is being used as a solvent, high-temperature distillation is a way to separate the carrier salt for reuse.

Molten salt reactor designs carry out fluoride volatility reprocessing continuously or at frequent intervals. The goal is to return actinides to the molten fuel mixture for eventual fission, while removing fission products that are neutron poisons, or that can be more securely stored outside the reactor core while awaiting eventual transfer to permanent storage.

Chloride Volatility and Solubility

Many of the elements that form volatile high-valence fluorides will also form volatile high-valence chlorides. Chlorination and distillation is another possible method for separation. The sequence of separation may differ usefully from the sequence for fluorides; for example, zirconium tetrachloride and tin tetrachloride have relatively low boiling points of 331 °C and 114.1 °C. Chlorination has even been proposed as a method for removing zirconium fuel cladding, instead of mechanical decladding.

Chlorides are likely to be easier than fluorides to later convert back to other compounds, such as oxides.

Chlorides remaining after volatilization may also be separated by solubility in water. Chlorides of alkaline elements like americium, curium, lanthanides, strontium, caesium are more soluble than those of uranium, neptunium, plutonium, and zirconium.

Radioanalytical Separations

In order to determine the distribution of radioactive metals for analytical purposes, Solvent Impregnated Resins (SIRs) can be used. SIRs are porous particles, which contain an extractant inside their pores. This approach avoids the liquid-liquid separation step required in conventional liquid-liquid extraction. For the preparation of SIRs for radioanalytical separations, organic Amberlite XAD-4 or XAD-7 can be used. Possible extractants are e.g. trihexyltetradecylphosphonium chloride(CYPHOS IL-101) or N,N0-dialkyl-N,N0-diphenylpyridine-2,6-dicarboxyamides (R-PDA; R = butyl, octy I, decyl, dodecyl).

Economics

The relative economics of reprocessing-waste disposal and interim storage-direct disposal has been the focus of much debate over the past ten years. Studies have modeled the total fuel cycle costs of a reprocessing-recycling system based on one-time recycling of plutonium in existing thermal reactors (as opposed to the proposed breeder reactor cycle) and compare this to the total costs of an open fuel cycle with direct disposal. The range of results produced by these studies is very wide, but all are agreed that under current (2005) economic conditions the reprocessing-recycle option is the more costly.

If reprocessing is undertaken only to reduce the radioactivity level of spent fuel it should be taken into account that spent nuclear fuel becomes less radioactive over time. After 40 years its radioactivity drops by 99.9%, though it still takes over a thousand years for the level of radioactivity to approach that of natural uranium. However the level of transuranic elements, including plutonium-239, remains high for over 100,000 years, so if not reused as nuclear fuel, then those elements need secure disposal because of nuclear proliferation reasons as well as radiation hazard.

On 25 October 2011 a commission of the Japanese Atomic Energy Commission revealed during a meeting calculations about the costs of recycling nuclear fuel for power generation. These costs could be twice the costs of direct geological disposal of spent fuel: the cost of extracting plutonium and handling spent fuel was estimated at 1.98 to 2.14 yen per kilowatt-hour of electricity generated. Discarding the spent fuel as waste would cost only 1 to 1.35 yen per kilowatt-hour.

In July 2004 Japanese newspapers reported that the Japanese Government had estimated the costs of disposing radioactive waste, contradicting claims four months earlier that no such estimates had been made. The cost of non-reprocessing options was estimated to be between a quarter and a third ($5.5–7.9 billion) of the cost of reprocessing ($24.7 billion). At the end of the year 2011 it became clear that Masaya Yasui, who had been director of the Nuclear Power Policy Planning Division in 2004, had instructed his subordinate in April 2004 to conceal the data. The fact that the data were deliberately concealed obliged the ministry to re-investigate the case and to reconsider whether to punish the officials involved.

List of Sites

Country	Reprocessing site	Fuel type	Procedure	Reprocessing capacity tU/yr	Commissioning or operating period
Belgium	Mol	LWR, MTR (Material test reactor)		80	1966–1974
China	intermediate pilot plant			60–100	1968-early 1970s
China	Plant 404			50	2004
Germany	Karlsruhe, WAK	LWR		35	1971–1990
France	Marcoule, UP 1	Military		1,200	1958-1997
France	Marcoule, CEA APM	FBR	PUREX DIA-MEX SANEX	6	1988–present
France	La Hague, UP 2	LWR	PUREX	900	1967–1974
France	La Hague, UP 2–400	LWR	PUREX	400	1976–1990
France	La Hague, UP 2–800	LWR	PUREX	800	1990
France	La Hague, UP 3	LWR	PUREX	800	1990
UK	Windscale	Magnox	BUTEX	1,000	1956–1962
UK	Sellafield, B205	Magnox	PUREX	1,500	1964
UK	Dounreay	FBR		8	1980
UK	THORP	AGR, LWR	PUREX	900	1994
Italy	Rotondella	Thorium		5	1968 (shutdown)
India	Trombay	Military	PUREX	60	1965

Country	Plant	Type	Process	Capacity	Years
India	Tarapur	PHWR		100	1982
India	Kalpakkam	PHWR and FBTR		100	1998
India	Tarapur	PHWR		100	2011
Japan	Tokaimura	LWR		210	1977-2006
Japan	Rokkasho	LWR		800	2005
Pakistan	New Labs, Rawalpindi	Military/Plutonium/Thorium		80	1982 – present
Pakistan	Khushab Nuclear Complex, Atomic City of Pakistan	HWR/Military/Tritium		22 kg	1986 – present
Russia	Mayak Plant B	Military		400	1948-196?
Russia	Mayak Plant BB, RT-1	LWR	PUREX + Np separation	400	1978
Russia	Zheleznogorsk (Krasnoyarsk-26), RT-2	VVER		800	under construction (2030)
USA, WA	Hanford Site	Military	bismuth phosphate, REDOX, PUREX		1944–1988
USA, SC	Savannah River Site	Military/LWR/HWR/Tritium	PUREX, REDOX, THOREX, Np separation	5000	1952–2002
USA, NY	West Valley	LWR	PUREX	300	1966–1972

High-level Radioactive Waste Management

High-level radioactive waste management concerns how radioactive materials created during production of nuclear power and nuclear weapons are dealt with. Radioactive waste contains a mixture of short-lived and long-lived nuclides, as well as non-radioactive nuclides. There was reported some 47,000 tonnes of high-level nuclear waste stored in the USA in 2002.

The most troublesome transuranic elements in spent fuel are neptunium-237 (half-life two million years) and plutonium-239 (half-life 24,000 years). Consequently, high-level radioactive waste requires sophisticated treatment and management to successfully isolate it from the biosphere. This usually necessitates treatment, followed by a long-term management strategy involving permanent storage, disposal or transformation of the waste into a non-toxic form. Radioactive decay follows the half-life rule, which means that the rate of decay is inversely proportional to the duration of decay. In other words, the radiation from a long-lived isotope like iodine-129 will be much less intense than that of short-lived isotope like iodine-131.

Spent nuclear fuel stored underwater and uncapped at the Hanford site in Washington, USA.

Governments around the world are considering a range of waste management and disposal options, usually involving deep-geologic placement, although there has been limited progress toward implementing long-term waste management solutions. This is partly because the timeframes in question when dealing with radioactive waste range from 10,000 to millions of years, according to studies based on the effect of estimated radiation doses.

Thus, Alfvén identified two fundamental prerequisites for effective management of high-level radioactive waste: (1) stable geological formations, and (2) stable human institutions over hundreds of thousands of years. As Alfvén suggests, no known human civilization has ever endured for so long, and no geologic formation of adequate size for a permanent radioactive waste repository has yet been discovered that has been stable for so long a period. Nevertheless, avoiding confronting the risks associated with managing radioactive wastes may create countervailing risks of greater magnitude. Radioactive waste management is an example of policy analysis that requires special attention to ethical concerns, examined in the light of uncertainty and *futurity*: consideration of 'the impacts of practices and technologies on future generations'.

There is a debate over what should constitute an acceptable scientific and engineering foundation for proceeding with radioactive waste disposal strategies. There are those who have argued, on the basis of complex geochemical simulation models, that relinquishing control over radioactive materials to geohydrologic processes at repository closure is an acceptable risk. They maintain that so-called "natural analogues" inhibit subterranean movement of radionuclides, making disposal of radioactive wastes in stable geologic formations unnecessary. However, existing models of these processes are empirically underdetermined: due to the subterranean nature of such processes in solid geologic formations, the accuracy of computer simulation models has not been verified by empirical observation, certainly not over periods of time equivalent to the lethal half-lives of high-level radioactive waste. On the other hand, some insist deep geologic repositories in stable geologic formations are necessary. National management plans of various countries display a variety of approaches to resolving this debate.

Researchers suggest that forecasts of health detriment for such long periods *should be examined critically*. Practical studies only consider up to 100 years as far as effective planning and cost evaluations are concerned. Long term behaviour of radioactive wastes remains a subject for ongoing research. Management strategies and implementation plans of several representative national governments are described below.

Geologic Disposal

The International Panel on Fissile Materials has said:

It is widely accepted that spent nuclear fuel and high-level reprocessing and plutonium wastes require well-designed storage for periods ranging from tens of thousands to a million years, to minimize releases of the contained radioactivity into the environment. Safeguards are also required to ensure that neither plutonium nor highly enriched uranium is diverted to weapon use. There is general agreement that placing spent nuclear fuel in repositories hundreds of meters below the surface would be safer than indefinite storage of spent fuel on the surface.

The process of selecting appropriate permanent repositories for high level waste and spent fuel is now under way in several countries with the first expected to be commissioned some time after 2017. The basic concept is to locate a large, stable geologic formation and use mining technology to excavate a tunnel, or large-bore tunnel boring machines (similar to those used to drill the Chunnel from England to France) to drill a shaft 500–1,000 meters below the surface where rooms or vaults can be excavated for disposal of high-level radioactive waste. The goal is to permanently isolate nuclear waste from the human environment. However, many people remain uncomfortable with the immediate stewardship cessation of this disposal system, suggesting perpetual management and monitoring would be more prudent.

Because some radioactive species have half-lives longer than one million years, even very low container leakage and radionuclide migration rates must be taken into account. Moreover, it may require more than one half-life until some nuclear materials lose enough radioactivity to no longer be lethal to living organisms. A 1983 review of the Swedish radioactive waste disposal program by the National Academy of Sciences found that country's estimate of several hundred thousand years—perhaps up to one million years—being necessary for waste isolation "fully justified."

The proposed land-based subductive waste disposal method would dispose of nuclear waste in a subduction zone accessed from land, and therefore is not prohibited by international agreement. This method has been described as a viable means of disposing of radioactive waste, and as a state-of-the-art nuclear waste disposal technology.

In nature, sixteen repositories were discovered at the Oklo mine in Gabon where natural nuclear fission reactions took place 1.7 billion years ago. The fission products in these natural formations were found to have moved less than 10 ft (3 m) over this period, though the lack of movement may be due more to retention in the uraninite structure than to insolubility and sorption from moving ground water; uraninite crystals are better preserved here than those in spent fuel rods because of a less complete nuclear reaction, so that reaction products would be less accessible to groundwater attack.

Materials for Geological Disposal

In order to store the high level radioactive waste in long-term geological depositories, specific waste forms need to be used which will allow the radioactivity to decay away while the materials retain their integrity for thousands of years. The materials being used can be broken down into a few classes: glass waste forms, ceramic waste forms, and nanostructured materials.

The glass forms include borosilicate glasses and phosphate glasses. Borosilicate nuclear waste glasses are used on an industrial scale to immobilize high level radioactive waste in many countries which are producers of nuclear energy or have nuclear weaponry. The glass waste forms have the advantage of being able to accommodate a wide variety of waste-stream compositions, they are easy to scale up to industrial processing, and they are stable against thermal, radiative, and chemical perturbations. These glasses function by binding radioactive elements to nonradioactive glass-forming elements. Phosphate glasses while not being used industrially have much lower dissolution rates than borosilicate glasses, which make them a more favorable option. However, no single phosphate material has the ability to accommodate all of the radioactive products so phosphate storage requires more reprocessing to separate the waste into distinct fractions. Both glasses have to be processed at elevated temperatures making them unusable for some of the more volatile radiotoxic elements.

The ceramic waste forms offer higher waste loadings than the glass options because ceramics have crystalline structure. Also, mineral analogues of the ceramic waste forms provide evidence for long term durability. Due to this fact and the fact that they can be processed at lower temperatures, ceramics are often considered the next generation in high level radioactive waste forms. Ceramic waste forms offer great potential, but a lot of research remains to be done.

National Management Plans

Finland, the United States and Sweden are the most advanced in developing a deep repository for high-level radioactive waste disposal. Countries vary in their plans on disposing used fuel directly or after reprocessing, with France and Japan having an extensive commitment to reprocessing. The country-specific status of high-level waste management plans are described below.

In many European countries (e.g., Britain, Finland, the Netherlands, Sweden and Switzerland) the risk or dose limit for a member of the public exposed to radiation from a future high-level nuclear waste facility is considerably more stringent than that suggested by the International Commission on Radiation Protection or proposed in the United States. European limits are often more stringent than the standard suggested in 1990 by the International Commission on Radiation Protection by a factor of 20, and more stringent by a factor of ten than the standard proposed by the U.S. Environmental Protection Agency (EPA) for Yucca Mountain nuclear waste repository for the first 10,000 years after closure. Moreover, the U.S. EPA's proposed standard for greater than 10,000 years is 250 times more permissive than the European limit.

The countries that have made the most progress towards a repository for high-level radioactive waste have typically started with public consultations and made voluntary siting a necessary condition. This consensus seeking approach is believed to have a greater chance of success than top-down modes of decision making, but the process is necessarily slow, and there is "inadequate experience around the world to know if it will succeed in all existing and aspiring nuclear nations".

Moreover, most communities do not want to host a nuclear waste repository as they are "concerned about their community becoming a de facto site for waste for thousands of years, the health and environmental consequences of an accident, and lower property values".

Asia

People's Republic of China

In the Peoples Republic of China, ten reactors provide about 2% of electricity and five more are under construction. China made a commitment to reprocessing in the 1980s; a pilot plant is under construction at Lanzhou, where a temporary spent fuel storage facility has been constructed. Geological disposal has been studied since 1985, and a permanent deep geological repository was required by law in 2003. Sites in Gansu Province near the Gobi desert in northwestern China are under investigation, with a final site expected to be selected by 2020, and actual disposal by about 2050.

Republic of China

In the Republic of China, nuclear waste storage facility was built at the Southern tip of Orchid Island in Taitung County, offshore of Taiwan Island. The facility was built in 1982 and it is owned and operated by Taipower. The facility receives nuclear waste from Taipower's current three nuclear power plants. However, due to the strong resistance from local community in the island, the nuclear waste has to be stored at the power plant facilities themselves.

India

Sixteen nuclear reactors produce about 3% of India's electricity, and seven more are under construction. Spent fuel is processed at facilities in Trombay near Mumbai, at Tarapur on the west coast north of Mumbai, and at Kalpakkam on the southeast coast of India. Plutonium will be used in a fast breeder reactor (under construction) to produce more fuel, and other waste vitrified at Tarapur and Trombay. Interim storage for 30 years is expected, with eventual disposal in a deep geological repository in crystalline rock near Kalpakkam.

Japan

In 2000, a Specified Radioactive Waste Final Disposal Act called for creation of a new organization to manage high level radioactive waste, and later that year the Nuclear Waste Management Organization of Japan (NUMO) was established under the jurisdiction of the Ministry of Economy, Trade and Industry. NUMO is responsible for selecting a permanent deep geological repository site, construction, operation and closure of the facility for waste emplacement by 2040. Site selection began in 2002 and application information was sent to 3,239 municipalities, but by 2006, no local government had volunteered to host the facility. Kōchi Prefecture showed interest in 2007, but its mayor resigned due to local opposition. In December 2013 the government decided to identify suitable candidate areas before approaching municipalities.

The head of the Science Council of Japan's expert panel has said Japan's seismic conditions makes it difficult to predict ground conditions over the necessary 100,000 years, so it will be impossible to convince the public of the safety of deep geological disposal.

Europe

Belgium

Belgium has seven nuclear reactors that provide about 52% of its electricity. Belgian spent nuclear fuel was initially sent for reprocessing in France. In 1993, reprocessing was suspended following a resolution of the Belgian parliament; spent fuel is since being stored on the sites of the nuclear power plants. The deep disposal of high-level radioactive waste (HLW) has been studied in Belgium for more than 30 years. Boom Clay is studied as a reference host formation for HLW disposal. The Hades underground research laboratory (URL) is located at −223 m in the Boom Formation at the Mol site. The Belgian URL is operated by the Euridice Economic Interest Group, a joint organisation between SCK•CEN, the Belgian Nuclear Research Centre which initiated the research on waste disposal in Belgium in the 1970s and 1980s and ONDRAF/NIRAS, the Belgian agency for radioactive waste management. In Belgium, the regulatory body in charge of guidance and licensing approval is the Federal Agency of Nuclear Control, created in 2001.

Finland

In 1983, the government decided to select a site for permanent repository by 2010. With four nuclear reactors providing 29% of its electricity, Finland in 1987 enacted a Nuclear Energy Act making the producers of radioactive waste responsible for its disposal, subject to requirements of its Radiation and Nuclear Safety Authority and an absolute veto given to local governments in which a proposed repository would be located. Producers of nuclear waste organized the company Posiva, with responsibility for site selection, construction and operation of a permanent repository. A 1994 amendment to the Act required final disposal of spent fuel in Finland, prohibiting the import or export of radioactive waste.

Environmental assessment of four sites occurred in 1997–98, Posiva chose the Olkiluoto site near two existing reactors, and the local government approved it in 2000. The Finnish Parliament approved a deep geologic repository there in igneous bedrock at a depth of about 500 meters in 2001. The repository concept is similar to the Swedish model, with containers to be clad in copper and buried below the water table beginning in 2020. An underground characterization facility, Onkalo spent nuclear fuel repository, was under construction at the site in 2012.

France

With 58 nuclear reactors contributing about 75% of its electricity, the highest percentage of any country, France has been reprocessing its spent reactor fuel since the introduction of nuclear power there. Some reprocessed plutonium is used to make fuel, but more is being produced than is being recycled as reactor fuel. France also reprocesses spent fuel for other countries, but the nuclear waste is returned to the country of origin. Radioactive waste from reprocessing French spent fuel is expected to be disposed of in a geological repository, pursuant to legislation enacted in 1991 that established a 15-year period for conducting radioactive waste management research. Under this legislation, partition and transmutation of long-lived elements, immobilization and conditioning processes, and long-term near surface storage are being investigated by the Commissariat à l'Energie Atomique (CEA). Disposal in deep geological formations is being studied by the French agency for radioactive waste management, L'Agence Nationale pour la Gestion des Déchets Radioactifs, in underground research labs.

Three sites were identified for possible deep geologic disposal in clay near the border of Meuse and Haute-Marne, near Gard, and at Vienne. In 1998 the government approved the Meuse/Haute Marne Underground Research Laboratory, a site near Meuse/Haute-Marne and dropped the others from further consideration. Legislation was proposed in 2006 to license a repository by 2015, with operations expected in 2025.

Germany

Nuclear waste policy in Germany is in flux. German planning for a permanent geologic repository began in 1974, focused on salt dome Gorleben, a salt mine near Gorleben about 100 kilometers northeast of Braunschweig. The site was announced in 1977 with plans for a reprocessing plant, spent fuel management, and permanent disposal facilities at a single site. Plans for the reprocessing plant were dropped in 1979. In 2000, the federal government and utilities agreed to suspend underground investigations for three to ten years, and the government committed to ending its use of nuclear power, closing one reactor in 2003.

Within days of the March 2011 Fukushima Daiichi nuclear disaster, Chancellor Angela Merkel "imposed a three-month moratorium on previously announced extensions for Germany's existing nuclear power plants, while shutting seven of the 17 reactors that had been operating since 1981". Protests continued and, on 29 May 2011, Merkel's government announced that it would close all of its nuclear power plants by 2022.

Meanwhile, electric utilities have been transporting spent fuel to interim storage facilities at Gorleben, Lubmin and Ahaus until temporary storage facilities can be built near reactor sites. Previously, spent fuel was sent to France or the United Kingdom for reprocessing, but this practice was ended in July 2005.

Russia

In Russia, the Ministry of Atomic Energy (Minatom) is responsible for 31 nuclear reactors which generate about 16% of its electricity. Minatom is also responsible for reprocessing and radioactive waste disposal, including over 25,000 tons of spent nuclear fuel in temporary storage in 2001.

Russia has a long history of reprocessing spent fuel for military purposes, and previously planned to reprocess imported spent fuel, possibly including some of the 33,000 metric tons of spent fuel accumulated at sites in other countries who received fuel from the U.S., which the U.S. originally pledged to take back, such as Brazil, the Czech Republic, India, Japan, Mexico, Slovenia, South Korea, Switzerland, Taiwan, and the European Union.

An Environmental Protection Act in 1991 prohibited importing radioactive material for long-term storage or burial in Russia, but controversial legislation to allow imports for permanent storage was passed by the Russian Parliament and signed by President

Putin in 2001. In the long term, the Russian plan is for deep geologic disposal. Most attention has been paid to locations where waste has accumulated in temporary storage at Mayak, near Chelyabinsk in the Ural Mountains, and in granite at Krasnoyarsk in Siberia.

Sweden

In Sweden, as of 2007 there are ten operating nuclear reactors that produce about 45% of its electricity. Two other reactors in Barsebäck were shut down in 1999 and 2005. When these reactors were built, it was expected their nuclear fuel would be reprocessed in a foreign country, and the reprocessing waste would not be returned to Sweden. Later, construction of a domestic reprocessing plant was contemplated, but has not been built.

Passage of the Stipulation Act of 1977 transferred responsibility for nuclear waste management from the government to the nuclear industry, requiring reactor operators to present an acceptable plan for waste management with "absolute safety" in order to obtain an operating license. In early 1980, after the Three Mile Island meltdown in the United States, a referendum was held on the future use of nuclear power in Sweden. In late 1980, after a three-question referendum produced mixed results, the Swedish Parliament decided to phase out existing reactors by 2010. In 2010, the Swedish government opened up for construction of new nuclear reactors. The new units can only be built at the existing nuclear power sites, Oskarshamn, Ringhals or Forsmark, and only to replace one of the existing reactors, that will have to be shut down for the new one to be able to start up.

The Swedish Nuclear Fuel and Waste Management Company. (Svensk Kärnbränslehantering AB, known as SKB) was created in 1980 and is responsible for final disposal of nuclear waste there. This includes operation of a monitored retrievable storage facility, the Central Interim Storage Facility for Spent Nuclear Fuel at Oskarshamn, about 150 miles south of Stockholm on the Baltic coast; transportation of spent fuel; and construction of a permanent repository. Swedish utilities store spent fuel at the reactor site for one year before transporting it to the facility at Oskarshamn, where it will be stored in excavated caverns filled with water for about 30 years before removal to a permanent repository.

Conceptual design of a permanent repository was determined by 1983, calling for placement of copper-clad iron canisters in granite bedrock about 500 metres underground, below the water table in what is known as the KBS-3 method. Space around the canisters will be filled with bentonite clay. After examining six possible locations for a permanent repository, three were nominated for further investigation, at Osthammar, Oskarshamn, and Tierp. On 3 June 2009, Swedish Nuclear Fuel and Waste Co. chose a location for a deep-level waste site at Östhammar, near Forsmark Nuclear Power plant. The application to build the repository was handed in by SKB 2011.

Switzerland

Switzerland has five nuclear reactors that provide about 43% of its electricity. Some Swiss spent nuclear fuel has been sent for reprocessing in France and the United Kingdom; most fuel is being stored without reprocessing. An industry-owned organization, ZWILAG, built and operates a central interim storage facility for spent nuclear fuel and high-level radioactive waste, and for conditioning low-level radioactive waste and for incinerating wastes. Other interim storage facilities predating ZWILAG continue to operate in Switzerland.

The Swiss program is considering options for the siting of a deep repository for high-level radioactive waste disposal, and for low & intermediate level wastes. Construction of a repository is not foreseen until well into this century. Research on sedimentary rock (especially Opalinus Clay) is carried out at the Swiss Mont Terri rock laboratory; the Grimsel Test Site, an older facility in crystalline rock is also still active.

United Kingdom

Great Britain has 19 operating reactors, producing about 20% of its electricity. It processes much of its spent fuel at Sellafield on the northwest coast across from Ireland, where nuclear waste is vitrified and sealed in stainless steel canisters for dry storage above ground for at least 50 years before eventual deep geologic disposal. Sellafield has a history of environmental and safety problems, including a fire in a nuclear plant in Windscale, and a significant incident in 2005 at the main reprocessing plant (THORP).

In 1982 the Nuclear Industry Radioactive Waste Management Executive (NIREX) was established with responsibility for disposing of long-lived nuclear waste and in 2006 a Committee on Radioactive Waste Management (CoRWM) of the Department of Environment, Food and Rural Affairs recommended geologic disposal 200–1,000 meters underground. NIREX developed a generic repository concept based on the Swedish model but has not yet selected a site. A Nuclear Decommissioning Authority is responsible for packaging waste from reprocessing and will eventually relieve British Nuclear Fuels Ltd. of responsibility for power reactors and the Sellafield reprocessing plant.

North America

Canada

The 18 operating nuclear power plants in Canada generated about 16% of its electricity in 2006. A national Nuclear Fuel Waste Act was enacted by the Canadian Parliament in 2002, requiring nuclear energy corporations to create a waste management organization to propose to the Government of Canada approaches for management of nuclear waste, and implementation of an approach subsequently selected by the government. The Act defined management as "long term management by means of storage or disposal, including handling, treatment, conditioning or transport for the purpose of storage or disposal."

The resulting Nuclear Waste Management Organization(NWMO) conducted an extensive three-year study and consultation with Canadians. In 2005, they recommended Adaptive Phased Management, an approach that emphasized both technical and management methods. The technical method included centralized isolation and containment of spent nuclear fuel in a deep geologic repository in a suitable rock formation, such as the granite of the Canadian Shield or Ordovician sedimentary rocks. Also recommended was a phased decision making process supported by a program of continuous learning, research and development.

In 2007, the Canadian government accepted this recommendation, and NWMO was tasked with implementing the recommendation. No specific timeframe was defined for the process. In 2009, the NWMO was designing the process for site selection; siting was expected to take 10 years or more.

United States

The Nuclear Waste Policy Act of 1982 established a timetable and procedure for constructing a permanent, underground repository for high-level radioactive waste by the mid-1990s, and provided for some temporary storage of waste, including spent fuel from 104 civilian nuclear reactors that produce about 19.4% of electricity there. The United States in April 2008 had about 56,000 metric tons of spent fuel and 20,000 canisters of solid defense-related waste, and this is expected to increase to 119,000 metric tons by 2035. The U.S. opted for Yucca Mountain nuclear waste repository, a final repository at Yucca Mountain in Nevada, but this project was widely opposed, with some of the main concerns being long distance transportation of waste from across the United States to this site, the possibility of accidents, and the uncertainty of success in isolating nuclear waste from the human environment in perpetuity. Yucca Mountain, with capacity for 70,000 metric tons of radioactive waste, was expected to open in 2017. However, the Obama Administration rejected use of the site in the 2009 United States Federal Budget proposal, which eliminated all funding except that needed to answer inquiries from the Nuclear Regulatory Commission, "while the Administration devises a new strategy toward nuclear waste disposal." On March 5, 2009, Energy Secretary Steven Chu told a Senate hearing "the Yucca Mountain site no longer was viewed as an option for storing reactor waste." Starting in 1999, military-generated nuclear waste is being entombed at the Waste Isolation Pilot Plant in New Mexico.

In a Presidential Memorandum dated January 29, 2010, President Obama established the Blue Ribbon Commission on America's Nuclear Future (the Commission). The Commission, composed of fifteen members, conducted an extensive two-year study of nuclear waste disposal, what is referred to as the "back end" of the nuclear energy process. The Commission established three subcommittees: Reactor and Fuel Cycle Technology, Transportation and Storage, and Disposal. On January 26, 2012, the Commission submitted its final report to Energy Secretary Steven Chu. In the Disposal Subcommittee's final report the Commission does not issue recommendations for a specific site but rather presents a comprehensive recommendation for disposal strategies. During their research the Commission visited Finland, France, Japan, Russia, Sweden, and the UK. In their final report the Commission put forth seven recommendations for developing a comprehensive strategy to pursue:

Recommendation #1

> The United States should undertake an integrated nuclear waste management program that leads to the timely development of one or more permanent deep geological facilities for the safe disposal of spent fuel and high-level nuclear waste.

Recommendation #2

> A new, single-purpose organization is needed to develop and implement a focused, integrated program for the transportation, storage, and disposal 1 of nuclear waste in the United States.

Recommendation #3

> Assured access to the balance in the Nuclear Waste Fund (NWF) and to the revenues generated by annual nuclear waste fee payments from utility ratepayers is absolutely essential and must be provided to the new nuclear waste management organization.

Recommendation #4

> A new approach is needed to site and develop nuclear waste facilities in the United States in the future. We believe that these processes are most likely to succeed if they are:

- Adaptive—in the sense that process itself is flexible and produces decisions that are responsive to new information and new technical, social, or political developments.

- Staged—in the sense that key decisions are revisited and modified as necessary along the way rather than being pre-determined in advance.

- Consent-based—in the sense that affected communities have an opportunity to decide whether to accept facility siting decisions and retain significant local control.

- Transparent—in the sense that all stakeholders have an opportunity to understand key decisions and engage in the process in a meaningful way.

- Standards- and science-based—in the sense that the public can have confidence that all facilities meet rigorous, objective, and consistently-applied standards of safety and environmental protection.

- Governed by partnership arrangements or legally-enforceable agreements with host states, tribes and local communities.

Recommendation #5

The current division of regulatory responsibilities for long-term repository performance between the NRC and the EPA is appropriate and should continue. The two agencies should develop new, site-independent safety standards in a formally coordinated joint process that actively engages and solicits input from all the relevant constituencies.

Recommendation #6

The roles, responsibilities, and authorities of local, state, and tribal governments (with respect to facility siting and other aspects of nuclear waste disposal) must be an element of the negotiation between the federal government and the other affected units of government in establishing a disposal facility. In addition to legally-binding agreements, as discussed in Recommendation #4, all affected levels of government (local, state, tribal, etc.) must have, at a minimum, a meaningful consultative role in all other important decisions. Additionally, states and tribes should retain—or where appropriate, be delegated—direct authority over aspects of regulation, permitting, and operations where oversight below the federal level can be exercised effectively and in a way that is helpful in protecting the interests and gaining the confidence of affected communities and citizens.

Recommendation #7

The Nuclear Waste Technical Review Board (NWTRB) should be retained as a valuable source of independent technical advice and review.

International Repository

Although Australia does not have any nuclear power reactors, Pangea Resources considered siting an international repository in the outback of South Australia or Western Australia in 1998, but this stimulated legislative opposition in both states and the Australian national Senate during the following year. Thereafter, Pangea ceased operations in Australia but reemerged as Pangea International Association, and in 2002 evolved into the Association for Regional and International Underground Storage with support from Belgium, Bulgaria, Hungary, Japan and Switzerland. A general concept for an international repository has been advanced by one of the principals in all three ventures. Russia has expressed interest in serving as a repository for other countries, but does not envision sponsorship or control by an international body or group of other countries. South Africa, Argentina and western China have also been mentioned as possible locations.

In the EU, COVRA is negotiating a European-wide waste disposal system with single disposal sites that can be used by several EU-countries. This EU-wide storage possibility is being researched under the SAPIERR-2 program.

References

- Attix, Frank (1986). Introduction to Radiological Physics and Radiation Dosimetry. New York: Wiley-VCH. pp. 2–15,468,474. ISBN 978-0-471-01146-0.

- Anderson, Mary; Woessner, William (1992). Applied Groundwater Modeling. San Diego, CA: Academic Press Inc. pp. 325–327. ISBN 0-12-059485-4.

- "The 2007 Recommendations of the International Commission on Radiological Protection". Annals of the ICRP. ICRP publication 103. 37 (2–4). 2007. ISBN 978-0-7020-3048-2.

- Cochran, Robert (1999). The Nuclear Fuel Cycle: Analysis and Management. La Grange Park, IL: American Nuclear Society. pp. 52–57. ISBN 0-89448-451-6.

- Hafemeister, David W. (2007). Physics of societal issues: calculations on national security, environment, and energy. Berlin: Springer. p. 187. ISBN 0387689095.

- Shipman, J.T.; Wison J.D.; Todd A. (2007). An Introduction to Physical Science (10 ed.). Cengage Learning. p. 279. ISBN 978-0-618-93596-3.

- OECD Nuclear Energy Agency (May 2007). Management of recyclable fissile and fertile materials. OECD Publishing. p. 34. ISBN 978-92-64-03255-2. Retrieved 22 March 2011.

- Vandenbosch, Robert; Vandenbosch, Susanne E. (2007). Nuclear waste stalemate. Salt Lake City: University of Utah Press. ISBN 0-87480-903-7.

- Cameron L. Tracy, Megan K. Dustin & Rodney C. Ewing, Policy: Reassess New Mexico's nuclear-waste repository, Nature, 13 January 2016.

- "Pyroprocessing Technologies: Recycling used nuclear fuel for a sustainable energy future" (PDF). Argonne National Laboratory. 2012. p. 7. Retrieved 6 June 2016.

Recycling: An Overview

The converting of waste material into useful material in order to decrease the utilization of fresh raw materials is recycling. Battery recycling, paint recycling and concrete recycling are some of the recycling techniques contemplated in this chapter. Industrial waste treatment is best understood in confluence with the major topics listed in the following chapter.

Recycling

Recycling is the process of converting waste materials into reusable objects to prevent waste of potentially useful materials, reduce the consumption of fresh raw materials, energy usage, air pollution (from incineration) and water pollution (from landfilling) by decreasing the need for "conventional" waste disposal and lowering greenhouse gas emissions compared to plastic production. Recycling is a key component of modern waste reduction and is the third component of the "Reduce, Reuse and Recycle" waste hierarchy.

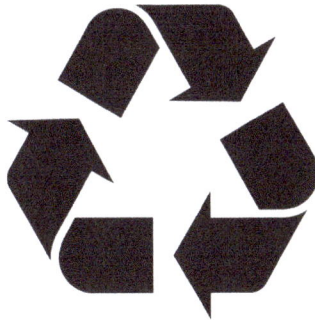

The three chasing arrows of the international recycling logo. It is sometimes accompanied by the text "reduce, reuse and recycle".

There are some ISO standards related to recycling such as ISO 15270:2008 for plastics waste and ISO 14001:2004 for environmental management control of recycling practice.

Recyclable materials include many kinds of glass, paper and cardboard, metal, plastic, tires, textiles and electronics. The composting or other reuse of biodegradable waste—such as food or garden waste—is also considered recycling. Materials to be recycled are either brought to a collection centre or picked up from the curbside, then sorted, cleaned and reprocessed into new materials destined for manufacturing.

In the strictest sense, recycling of a material would produce a fresh supply of the same material—for example, used office paper would be converted into new office paper, or used polystyrene foam into new polystyrene. However, this is often difficult or too expensive (compared with producing the same product from raw materials or other sources), so "recycling" of many products or materials involves their *reuse* in producing different materials (for example, paperboard) instead. Another form of recycling is the salvage of certain materials from complex products, either due to their intrinsic value (such as lead from car batteries, or gold from circuit boards), or due to their hazardous nature (e.g., removal and reuse of mercury from thermometers and thermostats).

History

Origins

Recycling has been a common practice for most of human history, with recorded advocates as far back as Plato in 400 BC. During periods when resources were scarce and hard to come by, archaeological studies of ancient waste dumps show less household waste (such as ash, broken tools and pottery)—implying more waste was being recycled in the absence of new material.

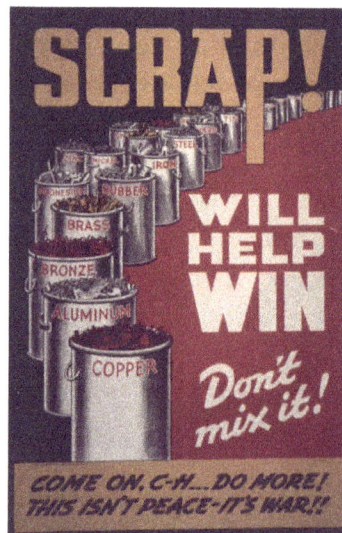

An American poster from World War II

In pre-industrial times, there is evidence of scrap bronze and other metals being collected in Europe and melted down for perpetual reuse. Paper recycling was first recorded in 1031, when Japanese shops sold repulped paper. In Britain dust and ash from wood and coal fires was collected by "dustmen" and downcycled as a base material used in brick making. The main driver for these types of recycling was the economic advantage of obtaining recycled feedstock instead of acquiring virgin material, as well as a lack of public waste removal in ever more densely populated areas. In 1813, Benjamin Law developed the process of turning rags into "shoddy" and "mungo" wool in

Batley, Yorkshire. This material combined recycled fibers with virgin wool. The West Yorkshire shoddy industry in towns such as Batley and Dewsbury, lasted from the early 19th century to at least 1914.

Industrialization spurred demand for affordable materials; aside from rags, ferrous scrap metals were coveted as they were cheaper to acquire than virgin ore. Railroads both purchased and sold scrap metal in the 19th century, and the growing steel and automobile industries purchased scrap in the early 20th century. Many secondary goods were collected, processed and sold by peddlers who scoured dumps and city streets for discarded machinery, pots, pans and other sources of metal. By World War I, thousands of such peddlers roamed the streets of American cities, taking advantage of market forces to recycle post-consumer materials back into industrial production.

Beverage bottles were recycled with a refundable deposit at some drink manufacturers in Great Britain and Ireland around 1800, notably Schweppes. An official recycling system with refundable deposits was established in Sweden for bottles in 1884 and aluminum beverage cans in 1982; the law led to a recycling rate for beverage containers of 84–99 percent depending on type, and a glass bottle can be refilled over 20 times on average.

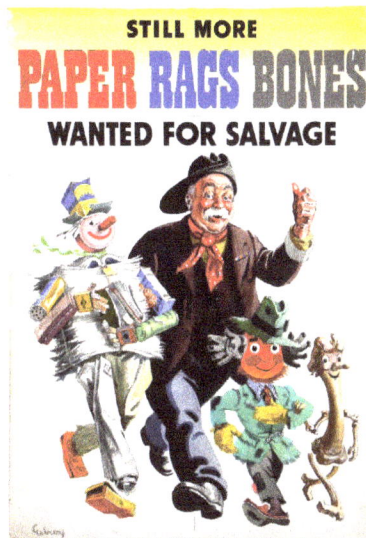

British poster from World War II

Wartime

New chemical industries created in the late 19th century both invented new materials (e.g. Bakelite (1907) and promised to transform valueless into valuable materials. Proverbially, you could not make a silk purse of a sow's ear—until the US firm Arhur D. Little published in 1921 "On the Making of Silk Purses from Sows' Ears", its research proving that when "chemistry puts on overalls and gets down to business . . .new values appear. New and better paths are opened to reach the goals desired."

Recycling was a highlight throughout World War II. During the war, financial constraints and significant material shortages due to war efforts made it necessary for countries to reuse goods and recycle materials. These resource shortages caused by the world wars, and other such world-changing occurrences, greatly encouraged recycling. The struggles of war claimed much of the material resources available, leaving little for the civilian population. It became necessary for most homes to recycle their waste, as recycling offered an extra source of materials allowing people to make the most of what was available to them. Recycling household materials meant more resources for war efforts and a better chance of victory. Massive government promotion campaigns were carried out in the home front during World War II in every country involved in the war, urging citizens to donate metals and conserve fiber, as a matter of patriotism.

Post-war

A considerable investment in recycling occurred in the 1970s, due to rising energy costs. Recycling aluminum uses only 5% of the energy required by virgin production; glass, paper and metals have less dramatic but very significant energy savings when recycled feedstock is used.

Although consumer electronics such as the television have been popular since the 1920s, recycling of them was almost unheard of until early 1991. The first electronic waste recycling scheme was implemented in Switzerland, beginning with collection of old refrigerators but gradually expanding to cover all devices. After these schemes were set up, many countries did not have the capacity to deal with the sheer quantity of e-waste they generated or its hazardous nature. They began to export the problem to developing countries without enforced environmental legislation. This is cheaper, as recycling computer monitors in the United States costs 10 times more than in China. Demand in Asia for electronic waste began to grow when scrap yards found that they could extract valuable substances such as copper, silver, iron, silicon, nickel and gold, during the recycling process. The 2000s saw a large increase in both the sale of electronic devices and their growth as a waste stream: in 2002, e-waste grew faster than any other type of waste in the EU. This caused investment in modern, automated facilities to cope with the influx of redundant appliances, especially after strict laws were implemented in 2003.

As of 2014, the European Union has about 50% of world share of the waste and recycling industries, with over 60,000 companies employing 500,000 persons, with a turnover of €24 billion. Countries have to reach recycling rates of at least 50%, while the lead countries are around 65% and the EU average is 39% as of 2013.

Legislation

Supply

For a recycling program to work, having a large, stable supply of recyclable material is crucial. Three legislative options have been used to create such a supply: mandatory re-cycling collection, container deposit legislation and refuse bans. Mandatory collection laws set recycling targets for cities to aim for, usually in the form that a certain percent-age of a material must be diverted from the city's waste stream by a target date. The city is then responsible for working to meet this target.

Container deposit legislation involves offering a refund for the return of certain con-tainers, typically glass, plastic and metal. When a product in such a container is pur-chased, a small surcharge is added to the price. This surcharge can be reclaimed by the consumer if the container is returned to a collection point. These programs have been very successful, often resulting in an 80 percent recycling rate. Despite such good re-sults, the shift in collection costs from local government to industry and consumers has created strong opposition to the creation of such programs in some areas. A variation on this is where the manufacturer bears responsibility for the recycling of their goods. In the European Union, the WEEE Directive requires producers of consumer electron-ics to reimburse the recyclers' costs.

An alternative way to increase supply of recyclates is to ban the disposal of certain ma-terials as waste, often including used oil, old batteries, tires and garden waste. One aim of this method is to create a viable economy for proper disposal of banned products. Care must be taken that enough of these recycling services exist, or such bans simply lead to increased illegal dumping.

Government-mandated Demand

Legislation has also been used to increase and maintain a demand for recycled ma-terials. Four methods of such legislation exist: minimum recycled content mandates, utilization rates, procurement policies and recycled product labeling.

Both minimum recycled content mandates and utilization rates increase demand directly by forcing manufacturers to include recycling in their operations. Content mandates spec-ify that a certain percentage of a new product must consist of recycled material. Utilization rates are a more flexible option: industries are permitted to meet the recycling targets at any point of their operation or even contract recycling out in exchange for tradeable credits. Opponents to both of these methods point to the large increase in reporting requirements they impose, and claim that they rob industry of necessary flexibility.

Governments have used their own purchasing power to increase recycling demand through what are called "procurement policies." These policies are either "set-asides," which reserve a certain amount of spending solely towards recycled products, or "price preference" programs which provide a larger budget when recycled items are pur-chased. Additional regulations can target specific cases: in the United States, for ex-ample, the Environmental Protection Agency mandates the purchase of oil, paper, tires and building insulation from recycled or re-refined sources whenever possible.

The final government regulation towards increased demand is recycled product labeling. When producers are required to label their packaging with amount of recycled material in the product (including the packaging), consumers are better able to make educated choices. Consumers with sufficient buying power can then choose more environmentally conscious options, prompt producers to increase the amount of recycled material in their products, and indirectly increase demand. Standardized recycling labeling can also have a positive effect on supply of recyclates if the labeling includes information on how and where the product can be recycled.

Recyclates

Glass recovered by crushing only one kind of beer bottle

Recyclate is a raw material that is sent to, and processed in a waste recycling plant or materials recovery facility which will be used to form new products. The material is collected in various methods and delivered to a facility where it undergoes re-manufacturing so that it can used in the production of new materials or products. For example, plastic bottles that are collected can be re-used and made into plastic pellets, a new product.

Quality of Recyclate

The quality of recyclates is recognized as one of the principal challenges that needs to be addressed for the success of a long-term vision of a green economy and achieving zero waste. Recyclate quality is generally referring to how much of the raw material is made up of target material compared to the amount of non-target material and other non-recyclable material. Only target material is likely to be recycled, so a higher amount of non-target and non-recyclable material will reduce the quantity of recycling product. A high proportion of non-target and non-recyclable material can make it more difficult for re-processors to achieve "high-quality" recycling. If the recyclate is of poor quality, it is more likely to end up being down-cycled or, in more extreme cases, sent to other recovery options or landfilled. For example, to facilitate the re-manufacturing of clear glass products there are tight restrictions for colored glass going into the re-melt process.

The quality of recyclate not only supports high quality recycling, but it can also deliver significant environmental benefits by reducing, reusing and keeping products out of landfills. High quality recycling can help support growth in the economy by maximizing the economic value of the waste material collected. Higher income levels from the sale of quality recyclates can return value which can be significant to local governments, households and businesses. Pursuing high quality recycling can also provide consumer and business confidence in the waste and resource management sector and may encourage investment in that sector.

There are many actions along the recycling supply chain that can influence and affect the material quality of recyclate. It begins with the waste producers who place non-target and non-recyclable wastes in recycling collection. This can affect the quality of final recyclate streams or require further efforts to discard those materials at later stages in the recycling process. The different collection systems can result in different levels of contamination. Depending on which materials are collected together, extra effort is required to sort this material back into separate streams and can significantly reduce the quality of the final product. Transportation and the compaction of materials can make it more difficult to separate material back into separate waste streams. Sorting facilities are not one hundred per cent effective in separating materials, despite improvements in technology and quality recyclate which can see a loss in recyclate quality. The storage of materials outside where the product can become wet can cause problems for re-processors. Reprocessing facilities may require further sorting steps to further reduce the amount of non-target and non-recyclable material. Each action along the recycling path plays a part in the quality of recyclate.

Quality Recyclate Action Plan (Scotland)

The Recyclate Quality Action Plan of Scotland sets out a number of proposed actions that the Scottish Government would like to take forward in order to drive up the quality of the materials being collected for recycling and sorted at materials recovery facilities before being exported or sold on to the reprocessing market.

The plan's objectives are to:

- Drive up the quality of recyclate.

- Deliver greater transparency about the quality of recyclate.

- Provide help to those contracting with materials recycling facilities to identify what is required of them

- Ensure compliance with the Waste (Scotland) regulations 2012.

- Stimulate a household market for quality recyclate.

- Address and reduce issues surrounding the Waste Shipment Regulations.

The plan focuses on three key areas, with fourteen actions which were identified to increase the quality of materials collected, sorted and presented to the processing market in Scotland.

The three areas of focus are:

1. Collection systems and input contamination

2. Sorting facilities – material sampling and transparency

3. Material quality benchmarking and standards

Recycling Consumer Waste

Collection

A number of different systems have been implemented to collect recyclates from the general waste stream. These systems lie along the spectrum of trade-off between public convenience and government ease and expense. The three main categories of collection are "drop-off centers," "buy-back centers" and "curbside collection."

A three-sided bin at a railway station in Germany, intended to separate paper *(left)* and plastic wrappings *(right)* from other waste *(back)*

Curbside Collection

Curbside collection encompasses many subtly different systems, which differ mostly on where in the process the recyclates are sorted and cleaned. The main categories are mixed waste collection, commingled recyclables and source separation. A waste collection vehicle generally picks up the waste.

A recycling truck collecting the contents of a recycling bin in Canberra, Australia

At one end of the spectrum is mixed waste collection, in which all recyclates are collected mixed in with the rest of the waste, and the desired material is then sorted out and cleaned at a central sorting facility. This results in a large amount of recyclable waste, paper especially, being too soiled to reprocess, but has advantages as well: the city need not pay for a separate collection of recyclates and no public education is needed. Any changes to which materials are recyclable is easy to accommodate as all sorting happens in a central location.

In a commingled or single-stream system, all recyclables for collection are mixed but kept separate from other waste. This greatly reduces the need for post-collection cleaning but does require public education on what materials are recyclable.

Source separation is the other extreme, where each material is cleaned and sorted prior to collection. This method requires the least post-collection sorting and produces the purest recyclates, but incurs additional operating costs for collection of each separate material. An extensive public education program is also required, which must be successful if recyclate contamination is to be avoided.

Source separation used to be the preferred method due to the high sorting costs incurred by commingled (mixed waste) collection. Advances in sorting technology, however, have lowered this overhead substantially—many areas which had developed source separation programs have since switched to co-mingled collection.

Buy-back Centers

Buy-back centers differ in that the cleaned recyclates are purchased, thus providing a clear incentive for use and creating a stable supply. The post-processed material can then be sold on, hopefully creating a profit. Unfortunately, government subsidies are necessary to make buy-back centres a viable enterprise, as according to the U.S. National Waste & Recycling Association, it costs on average US$50 to process a ton of material, which can only be resold for US$30.

Drop-off Centers

Drop-off centers require the waste producer to carry the recyclates to a central location, either an installed or mobile collection station or the reprocessing plant itself. They are the easiest type of collection to establish, but suffer from low and unpredictable throughput.

Distributed Recycling

For some waste materials such as plastic, recent technical devices called recyclebots enable a form of distributed recycling. Preliminary life-cycle analysis (LCA) indicates that such distributed recycling of HDPE to make filament of 3-D printers in rural regions is energetically favorable to either using virgin resin or conventional recycling processes because of reductions in transportation energy.

Sorting

Once commingled recyclates are collected and delivered to a central collection facility, the different types of materials must be sorted. This is done in a series of stages, many of which involve automated processes such that a truckload of material can be fully sorted in less than an hour. Some plants can now sort the materials automatically, known as single-stream recycling. In plants, a variety of materials are sorted such as paper, different types of plastics, glass, metals, food scraps and most types of batteries. A 30 percent increase in recycling rates has been seen in the areas where these plants exist.

Initially, the commingled recyclates are removed from the collection vehicle and placed on a conveyor belt spread out in a single layer. Large pieces of corrugated fiberboard and plastic bags are removed by hand at this stage, as they can cause later machinery to jam.

Early sorting of recyclable materials: glass and plastic bottles in Poland

Next, automated machinery such as disk screens and air classifiers separate the recyclates by weight, splitting lighter paper and plastic from heavier glass and metal. Cardboard is removed from the mixed paper and the most common types of plastic, PET (#1) and HDPE (#2), are collected. This separation is usually done by hand but has become automated in some sorting centers: a spectroscopic scanner is used to differentiate between different types of paper and plastic based on the absorbed wavelengths, and subsequently divert each material into the proper collection channel.

A recycling point in New Byth, Scotland, with separate containers for paper, plastics and differently colored glass

Strong magnets are used to separate out ferrous metals, such as iron, steel and tin cans. Non-ferrous metals are ejected by magnetic eddy currents in which a rotating magnetic field induces an electric current around the aluminum cans, which in turn creates a magnetic eddy current inside the cans. This magnetic eddy current is repulsed by a large magnetic field, and the cans are ejected from the rest of the recyclate stream.

Finally, glass is sorted according to its color: brown, amber, green or clear. It may either be sorted by hand, or via an automated machine that uses colored filters to detect different colors. Glass fragments smaller than 10 millimetres (0.39 in) across cannot be sorted automatically, and are mixed together as "glass fines."

This process of recycling as well as reusing the recycled material has proven advantageous because it reduces amount of waste sent to landfills, conserves natural resources, saves energy, reduces greenhouse gas emissions and helps create new jobs. Recycled materials can also be converted into new products that can be consumed again, such as paper, plastic and glass.

The City and County of San Francisco's Department of the Environment is attempting to achieve a citywide goal of Zero Waste by 2020. San Francisco's refuse hauler, Recology, operates an effective recyclables sorting facility in San Francisco, which helped San Francisco reach a record-breaking diversion rate of 80%.

Rinsing

Food packaging should no longer contain any organic matter (organic matter, if any, needs to be placed in a biodegradable waste bin or be buried in a garden). Since no trace of biodegradable material is best kept in the packaging before placing it in a trash bag, some packaging also needs to be rinsed.

Recycling Industrial Waste

Mounds of shredded rubber tires are ready for processing

Although many government programs are concentrated on recycling at home, a 64% of waste in the United Kingdom is generated by industry. The focus of many recycling programs done by industry is the cost–effectiveness of recycling. The ubiquitous nature of cardboard packaging makes cardboard a commonly recycled waste product by companies that deal heavily in packaged goods, like retail stores, warehouses and distributors of goods. Other industries deal in niche or specialized products, depending on the nature of the waste materials that are present.

The glass, lumber, wood pulp and paper manufacturers all deal directly in commonly recycled materials; however, old rubber tires may be collected and recycled by independent tire dealers for a profit.

Levels of metals recycling are generally low. In 2010, the International Resource Panel, hosted by the United Nations Environment Programme (UNEP) published reports on metal stocks that exist within society and their recycling rates. The Panel reported that the increase in the use of metals during the 20th and into the 21st century has led to a substantial shift in metal stocks from below ground to use in applications within society above ground. For example, the in-use stock of copper in the USA grew from 73 to 238 kg per capita between 1932 and 1999.

The report authors observed that, as metals are inherently recyclable, the metal stocks in society can serve as huge mines above ground (the term "urban mining" has been coined with this idea in mind). However, they found that the recycling rates of many metals are very low. The report warned that the recycling rates of some rare metals used in applications such as mobile phones, battery packs for hybrid cars and fuel cells, are so low that unless future end-of-life recycling rates are dramatically stepped up these critical metals will become unavailable for use in modern technology.

The military recycles some metals. The U.S. Navy's Ship Disposal Program uses ship breaking to reclaim the steel of old vessels. Ships may also be sunk to create an artificial reef. Uranium is a very dense metal that has qualities superior to lead and titanium for many military and industrial uses. The uranium left over from processing it into nuclear weapons and fuel for nuclear reactors is called depleted uranium, and it is used by all branches of the U.S. military use for armour-piercing shells and shielding.

The construction industry may recycle concrete and old road surface pavement, selling their waste materials for profit.

Some industries, like the renewable energy industry and solar photovoltaic technology in particular, are being proactive in setting up recycling policies even before there is considerable volume to their waste streams, anticipating future demand during their rapid growth.

Recycling of plastics is more difficult, as most programs are not able to reach the necessary level of quality. Recycling of PVC often results in downcycling of the material, which means only products of lower quality standard can be made with the recycled material. A new approach which allows an equal level of quality is the Vinyloop process. It was used after the London Olympics 2012 to fulfill the PVC Policy.

E-waste Recycling

E-waste is a growing problem, accounting for 20-50 million metric tons of global waste per year according to the EPA. It is also the fastest growing waste stream in the EU. Many recyclers do not recycle e-waste responsibly. After the cargo barge Khian Sea dumped 14,000 metric tons of toxic ash in Haiti, the Basel Convention was formed to stem the flow of hazardous substances into poorer countries. They created the e-Stewards certification to ensure that recyclers are held to the highest standards for environmental responsibility and to help consumers identify responsible recyclers. This works alongside other prominent legislation, such as the Waste Electrical and Electronic Equipment Directive of the EU the United States National Computer Recycling Act, to prevent poisonous chemicals from entering waterways and the atmosphere.

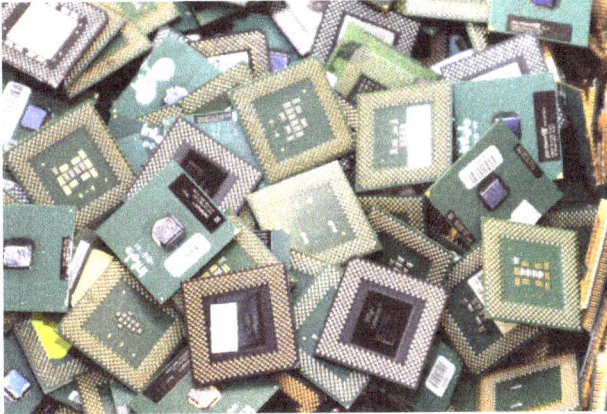
Microprocessors retrieved from waste stream

In the recycling process, television sets, monitors, cell phones and computers are typically tested for reuse and repaired. If broken, they may be disassembled for parts still having high value if labor is cheap enough. Other e-waste is shredded to pieces roughly 10 centimetres (3.9 in) in size, and manually checked to separate out toxic batteries and capacitors which contain poisonous metals. The remaining pieces are further shredded to 10 millimetres (0.39 in) particles and passed under a magnet to remove ferrous metals. An eddy current ejects non-ferrous metals, which are sorted by density either by a centrifuge or vibrating plates. Precious metals can be dissolved in acid, sorted, and smelted into ingots. The remaining glass and plastic fractions are separated by density and sold to re-processors. Television sets and monitors must be manually disassembled to remove lead from CRTs or the mercury backlight from LCDs.

Plastic Recycling

Plastic recycling is the process of recovering scrap or waste plastic and reprocessing the material into useful products, sometimes completely different in form from their original state. For instance, this could mean melting down soft drink bottles and then casting them as plastic chairs and tables.

A container for recycling used plastic spoons into material for 3D printing

Physical Recycling

Some plastics are remelted to form new plastic objects; for example, PET water bottles can be converted into polyester destined for clothing. A disadvantage of this type of recycling is that the molecular weight of the polymer can change further and the levels of unwanted substances in the plastic can increase with each remelt.

Chemical Recycling

For some polymers, it is possible to convert them back into monomers, for example PET can be treated with an alcohol and a catalyst to form a dialkyl terephthalate. The terephthalate diester can be used with ethylene glycol to form a new polyester polymer, thus making it possible to use the pure polymer again.

Waste Plastic Pyrolysis to Fuel Oil

Another process involves conversion of assorted polymers into petroleum by a much less precise thermal depolymerization process. Such a process would be able to accept almost any polymer or mix of polymers, including thermoset materials such as vulcanized rubber tires and the biopolymers in feathers and other agricultural waste. Like natural petroleum, the chemicals produced can be used as fuels or as feedstock. A RESEM Technology plant of this type in Carthage, Missouri, USA, uses turkey waste as input material. Gasification is a similar process, but is not technically recycling since polymers are not likely to become the result. Plastic Pyrolysis can convert petroleum based waste streams such as plastics into quality fuels, carbons. Given below is the list of suitable plastic raw materials for pyrolysis:

- Mixed plastic (HDPE, LDPE, PE, PP, Nylon, Teflon, PS, ABS, FRP, etc.)

- Mixed waste plastic from waste paper mill

- Multi-layered plastic

Recycling Codes

In order to meet recyclers' needs while providing manufacturers a consistent, uniform system, a coding system was developed. The recycling code for plastics was introduced in 1988 by the plastics industry through the Society of the Plastics Industry. Because municipal recycling programs traditionally have targeted packaging—primarily bottles and containers—the resin coding system offered a means of identifying the resin content of bottles and containers commonly found in the residential waste stream.

Recycling codes on products

Plastic products are printed with numbers 1–7 depending on the type of resin. Type 1 (polyethylene terephthalate) is commonly found in soft drink and water bottles. Type 2 (high-density polyethylene) is found in most hard plastics such as milk jugs, laundry detergent bottles and some dishware. Type 3 (polyvinyl chloride) includes items such as shampoo bottles, shower curtains, hula hoops, credit cards, wire jacketing, medical equipment, siding and piping. Type 4 (low-density polyethylene) is found in shopping bags, squeezable bottles, tote bags, clothing, furniture and carpet. Type 5 is polypropylene and makes up syrup bottles, straws, Tupperware and some automotive parts. Type 6 is polystyrene and makes up meat trays, egg cartons, clamshell containers and compact disc cases. Type 7 includes all other plastics such as bulletproof materials, 3- and 5-gallon water bottles and sunglasses. Having a recycling code or the chasing arrows logo on a material is not an automatic indicator that a material is recyclable but rather an explanation of what the material is. Types 1 and 2 are the most commonly recycled.

Economic Impact

Critics dispute the net economic and environmental benefits of recycling over its costs, and suggest that proponents of recycling often make matters worse and suffer from confirmation bias. Specifically, critics argue that the costs and energy used in collection and transportation detract from (and outweigh) the costs and energy saved in the production process; also that the jobs produced by the recycling industry can be a poor trade for the jobs lost in logging, mining, and other industries associated with production; and that materials such as paper pulp can only be recycled a few times before material degradation prevents further recycling.

The National Waste and Recycling Association (NWRA), reported in May 2015, that recycling and waste made a $6.7 billion economic impact in Ohio, U.S., and employed 14,000 people.

Cost–Benefit Analysis

Environmental effects of recycling		
Material	**Energy savings**	**Air pollution savings**
Aluminium	95%	95%
Cardboard	24%	—
Glass	5–30%	20%
Paper	40%	73%
Plastics	70%	—
Steel	60%	—

There is some debate over whether recycling is economically efficient. It is said that dumping 10,000 tons of waste in a landfill creates six jobs, while recycling 10,000 tons of waste can create over 36 jobs. However, the cost effectiveness of creating the additional jobs remains unproven. According to the U.S. Recycling Economic Informational Study, there are over 50,000 recycling establishments that have created over a million jobs in the US. Two years after New York City declared that implementing recycling programs would be "a drain on the city," New York City leaders realized that an efficient recycling system could save the city over $20 million. Municipalities often see fiscal benefits from implementing recycling programs, largely due to the reduced landfill costs. A study conducted by the Technical University of Denmark according to the Economist found that in 83 percent of cases, recycling is the most efficient method to dispose of household waste. However, a 2004 assessment by the Danish Environmental Assessment Institute concluded that incineration was the most effective method for disposing of drink containers, even aluminium ones.

Fiscal efficiency is separate from economic efficiency. Economic analysis of recycling does not include what economists call externalities, which are unpriced costs and benefits that accrue to individuals outside of private transactions. Examples include: decreased air pollution and greenhouse gases from incineration, reduced hazardous waste leaching from landfills, reduced energy consumption, and reduced waste and resource consumption, which leads to a reduction in environmentally damaging mining and timber activity. About 4,000 minerals are known, of these only a few hundred minerals in the world are relatively common. Known reserves of phosphorus will be exhausted within the next 100 years at current rates of usage. Without mechanisms such as taxes or subsidies to internalize externalities, businesses will ignore them despite the costs imposed on society. To make such nonfiscal benefits economically relevant, advocates have pushed for legislative action to increase the demand for recycled materials. The United States Environmental Protection Agency (EPA) has concluded in favor of recycling, saying that recycling efforts reduced the country's carbon emissions by a net 49 million metric tonnes in 2005. In the United Kingdom, the Waste and Resources Action Programme stated that Great Britain's recycling efforts reduce CO_2 emissions by 10–15 million tonnes a year. Recycling is more efficient in densely populated areas, as there are economies of scale involved.

Wrecked automobiles gathered for smelting

Certain requirements must be met for recycling to be economically feasible and environmentally effective. These include an adequate source of recyclates, a system to extract those recyclates from the waste stream, a nearby factory capable of reprocessing the recyclates, and a potential demand for the recycled products. These last two requirements are often overlooked—without both an industrial market for production using the collected materials and a consumer market for the manufactured goods, recycling is incomplete and in fact only "collection".

Free-market economist Julian Simon remarked "There are three ways society can organize waste disposal: (a) commanding,(b) guiding by tax and subsidy, and (c) leaving it to the individual and the market". These principles appear to divide economic thinkers today.

Frank Ackerman favours a high level of government intervention to provide recycling services. He believes that recycling's benefit cannot be effectively quantified by traditional *laissez-faire* economics. Allen Hershkowitz supports intervention, saying that it is a public service equal to education and policing. He argues that manufacturers should shoulder more of the burden of waste disposal.

Paul Calcott and Margaret Walls advocate the second option. A deposit refund scheme and a small refuse charge would encourage recycling but not at the expense of fly-tipping. Thomas C. Kinnaman concludes that a landfill tax would force consumers, companies and councils to recycle more.

Most free-market thinkers detest subsidy and intervention because they waste resources. Terry Anderson and Donald Leal think that all recycling programmes should be privately operated, and therefore would only operate if the money saved by recycling exceeds its costs. Daniel K. Benjamin argues that it wastes people's resources and lowers the wealth of a population.

Trade in Recyclates

Certain countries trade in unprocessed recyclates. Some have complained that the ultimate fate of recyclates sold to another country is unknown and they may end up in landfills instead of reprocessed. According to one report, in America, 50–80 percent of computers destined for recycling are actually not recycled. There are reports of illegal-waste imports to China being dismantled and recycled solely for monetary gain, without consideration for workers' health or environmental damage. Although the Chinese government has banned these practices, it has not been able to eradicate them. In 2008, the prices of recyclable waste plummeted before rebounding in 2009. Cardboard averaged about £53/tonne from 2004–2008, dropped to £19/tonne, and then went up to £59/tonne in May 2009. PET plastic averaged about £156/tonne, dropped to £75/tonne and then moved up to £195/tonne in May 2009.

Certain regions have difficulty using or exporting as much of a material as they recycle. This problem is most prevalent with glass: both Britain and the U.S. import large quantities of wine bottled in green glass. Though much of this glass is sent to be recycled, outside the American Midwest there is not enough wine production to use all of the reprocessed material. The extra must be downcycled into building materials or re-inserted into the regular waste stream.

Similarly, the northwestern United States has difficulty finding markets for recycled newspaper, given the large number of pulp mills in the region as well as the proximity to Asian markets. In other areas of the U.S., however, demand for used newsprint has seen wide fluctuation.

In some U.S. states, a program called RecycleBank pays people to recycle, receiving money from local municipalities for the reduction in landfill space which must be purchased. It uses a single stream process in which all material is automatically sorted.

Criticisms and Responses

Much of the difficulty inherent in recycling comes from the fact that most products are not designed with recycling in mind. The concept of sustainable design aims to solve this problem, and was laid out in the book *Cradle to Cradle: Remaking the Way We Make Things* by architect William McDonough and chemist Michael Braungart. They suggest that every product (and all packaging they require) should have a complete "closed-loop" cycle mapped out for each component—a way in which every component will either return to the natural ecosystem through biodegradation or be recycled indefinitely.

Complete recycling is impossible from a practical standpoint. In summary, substitution and recycling strategies only delay the depletion of non-renewable stocks and therefore may buy time in the transition to true or strong sustainability, which ultimately is only guaranteed in an economy based on renewable resources.

—M. H. Huesemann, 2003

While recycling diverts waste from entering directly into landfill sites, current recycling misses the dissipative components. Complete recycling is impracticable as highly dispersed wastes become so diluted that the energy needed for their recovery becomes increasingly excessive. "For example, how will it ever be possible to recycle the numerous chlorinated organic hydrocarbons that have bioaccumulated in animal and human tissues across the globe, the copper dispersed in fungicides, the lead in widely applied paints, or the zinc oxides present in the finely dispersed rubber powder that is abraded from automobile tires?"

As with environmental economics, care must be taken to ensure a complete view of the costs and benefits involved. For example, paperboard packaging for food products is more easily recycled than most plastic, but is heavier to ship and may result in more waste from spoilage.

Energy and Material Flows

Bales of crushed steel ready for transport to the smelter

The amount of energy saved through recycling depends upon the material being recycled and the type of energy accounting that is used. Correct accounting for this saved energy can be accomplished with life-cycle analysis using real energy values. In addition, exergy, which is a measure of useful energy can be used. In general, it takes far less energy to produce a unit mass of recycled materials than it does to make the same mass of virgin materials.

Some scholars use emergy (spelled with an m) analysis, for example, budgets for the amount of energy of one kind (exergy) that is required to make or transform things into another kind of product or service. Emergy calculations take into account economics which can alter pure physics based results. Using emergy life-cycle analysis researchers have concluded that materials with large refining costs have the greatest potential for high recycle benefits. Moreover, the highest emergy efficiency accrues from systems geared toward material recycling, where materials are engineered to recycle back into their original form and purpose, followed by adaptive reuse systems where the materials are recycled into a different kind of product, and then by-product reuse systems where parts of the products are used to make an entirely different product.

The Energy Information Administration (EIA) states on its website that "a paper mill uses 40 percent less energy to make paper from recycled paper than it does to make paper from fresh lumber." Some critics argue that it takes more energy to produce recycled products than it does to dispose of them in traditional landfill methods, since the curbside collection of recyclables often requires a second waste truck. However, recycling proponents point out that a second timber or logging truck is eliminated when paper is collected for recycling, so the net energy consumption is the same. An Emergy life-cycle analysis on recycling revealed that fly ash, aluminum, recycled concrete aggregate, recycled plastic, and steel yield higher efficiency ratios, whereas the recycling of lumber generates the lowest recycle benefit ratio. Hence, the specific nature of the recycling process, the methods used to analyse the process, and the products involved affect the energy savings budgets.

It is difficult to determine the amount of energy consumed or produced in waste disposal processes in broader ecological terms, where causal relations dissipate into complex networks of material and energy flow. For example, "cities do not follow all the strategies of ecosystem development. Biogeochemical paths become fairly straight relative to wild ecosystems, with very reduced recycling, resulting in large flows of waste and low total energy efficiencies. By contrast, in wild ecosystems, one population's wastes are another population's resources, and succession results in efficient exploitation of available resources. However, even modernized cities may still be in the earliest stages of a succession that may take centuries or millennia to complete." How much energy is used in recycling also depends on the type of material being recycled and the process used to do so. Aluminium is generally agreed to use far less energy when recycled rather than being produced from scratch. The EPA states that "recycling aluminum cans, for example, saves 95 percent of the energy required to make the same amount of aluminum from its virgin source, bauxite." In 2009 more than half of all aluminium cans produced came from recycled aluminium.

Every year, millions of tons of materials are being exploited from the earth's crust, and processed into consumer and capital goods. After decades to centuries, most of these materials are "lost". With the exception of some pieces of art or religious relics, they are no longer engaged in the consumption process. Where are they? Recycling is only an intermediate solution for such materials, although it does prolong the residence time in the anthroposphere. For thermodynamic reasons, however, recycling cannot prevent the final need for an ultimate sink.

— P. H. Brunner

Economist Steven Landsburg has suggested that the sole benefit of reducing landfill space is trumped by the energy needed and resulting pollution from the recycling process. Others, however, have calculated through life-cycle assessment that producing recycled paper uses less energy and water than harvesting, pulping, processing, and transporting virgin trees. When less recycled paper is used, additional energy is needed to create and maintain farmed forests until these forests are as self-sustainable as virgin forests.

Other studies have shown that recycling in itself is inefficient to perform the "decoupling" of economic development from the depletion of non-renewable raw materials that is necessary for sustainable development. The international transportation or recycle material flows through "… different trade networks of the three countries result in different flows, decay rates, and potential recycling returns." As global consumption of a natural resources grows, its depletion is inevitable. The best recycling can do is to delay, complete closure of material loops to achieve 100 percent recycling of nonrenewables is impossible as micro-trace materials dissipate into the environment causing severe damage to the planet's ecosystems. Historically, this was identified as the metabolic rift by Karl Marx, who identified the unequal exchange rate between energy and nutrients flowing from rural areas to feed urban cities that create effluent wastes degrading the planet's ecological capital, such as loss in soil nutrient production. Energy conservation also leads to what is known as Jevon's paradox, where improvements in energy efficiency lowers the cost of production and leads to a rebound effect where rates of consumption and economic growth increases.

A shop in New York only sells items recycled from demolished buildings

Costs

The amount of money actually saved through recycling depends on the efficiency of the recycling program used to do it. The Institute for Local Self-Reliance argues that the cost of recycling depends on various factors, such as landfill fees and the amount of disposal that the community recycles. It states that communities begin to save money when they treat recycling as a replacement for their traditional waste system rather than an add-on to it and by "redesigning their collection schedules and/or trucks."

In some cases, the cost of recyclable materials also exceeds the cost of raw materials. Virgin plastic resin costs 40 percent less than recycled resin. Additionally, a United States Environmental Protection Agency (EPA) study that tracked the price of clear glass from July 15 to August 2, 1991, found that the average cost per ton ranged from $40 to $60, while a USGS report shows that the cost per ton of raw silica sand from years 1993 to 1997 fell between $17.33 and $18.10.

In 1996 and 2015 articles for *The New York Times*, John Tierney argued that it costs more money to recycle the trash of New York City than it does to dispose of it in a landfill. Tierney argued that the recycling process employs people to do the additional waste disposal, sorting, inspecting, and many fees are often charged because the processing costs used to make the end product are often more than the profit from its sale. Tierney also referenced a study conducted by the Solid Waste Association of North America (SWANA) that found in the six communities involved in the study, "all but one of the curbside recycling programs, and all the composting operations and waste-to-energy incinerators, increased the cost of waste disposal."

Tierney also points out that "the prices paid for scrap materials are a measure of their environmental value as recyclables. Scrap aluminum fetches a high price because recycling it consumes so much less energy than manufacturing new aluminum."

However, comparing the market cost of recyclable material with the cost of new raw materials ignores economic externalities—the costs that are currently not counted by the market. Creating a new piece of plastic, for instance, may cause more pollution and be less sustainable than recycling a similar piece of plastic, but these factors will not be counted in market cost. A life cycle assessment can be used to determine the levels of externalities and decide whether the recycling may be worthwhile despite unfavorable market costs. Alternatively, legal means (such as a carbon tax) can be used to bring externalities into the market, so that the market cost of the material becomes close to the true cost.

In a 2007 article, Michael Munger, chairman of political science at Duke University, wrote that "if recycling is more expensive than using new materials, it can't possibly be efficient.... There is a simple test for determining whether something is a resource... or just garbage... If someone will pay you for the item, it's a resource.... But if you have to pay someone to take the item away,... then the item is garbage."

In a 2002 article for The Heartland Institute, Jerry Taylor, director of natural resource studies at the Cato Institute, wrote, "If it costs X to deliver newly manufactured plastic to the market, for example, but it costs 10X to deliver reused plastic to the market, we can conclude the resources required to recycle plastic are 10 times more scarce than the resources required to make plastic from scratch. And because recycling is supposed to be about the conservation of resources, mandating recycling under those circumstances will do more harm than good."

Working Conditions

The recycling of waste electrical and electronic equipment in India and China generates a significant amount of pollution. Informal recycling in an underground economy of these countries has generated an environmental and health disaster. High levels of lead (Pb), polybrominated diphenylethers (PBDEs), polychlorinated dioxins and furans, as well as polybrominated dioxins and furans (PCDD/Fs and PBDD/Fs) concentrated in the air, bottom ash, dust, soil, water and sediments in areas surrounding recycling sites. Critics also argue that while recycling may create jobs, they are often jobs with low wages and terrible working conditions. These jobs are sometimes considered to be make-work jobs that don't produce as much as the cost of wages to pay for those jobs. In areas without many environmental regulations and/or worker protections, jobs involved in recycling such as ship breaking can result in deplorable conditions for both workers and the surrounding communities.

People in Brazil who earn their living by collecting and sorting garbage and selling them for recycling

Environmental Impact

Economist Steven Landsburg, author of a paper entitled "Why I Am Not an Environmentalist," claimed that paper recycling actually reduces tree populations. He argues that because paper companies have incentives to replenish their forests, large demands for paper lead to large forests, while reduced demand for paper leads to fewer "farmed" forests.

When foresting companies cut down trees, more are planted in their place. Most paper comes from pulp forests grown specifically for paper production. Many environmentalists point out, however, that "farmed" forests are inferior to virgin forests in several ways. Farmed forests are not able to fix the soil as quickly as virgin forests, causing widespread soil erosion and often requiring large amounts of fertilizer to maintain while containing little tree and wild-life biodiversity compared to virgin forests. Also, the new trees planted are not as big as the trees that were cut down, and the argument that there will be "more trees" is not compelling to forestry advocates when they are counting saplings.

In particular, wood from tropical rainforests is rarely harvested for paper because of their heterogeneity. According to the United Nations Framework Convention on Climate Change secretariat, the overwhelming direct cause of deforestation is subsistence farming (48% of deforestation) and commercial agriculture (32%), which is linked to food, not paper production.

Possible Income Loss and Social Costs

In some countries, recycling is performed by the entrepreneurial poor such as the karung guni, zabbaleen, the rag-and-bone man, waste picker, and junk man. With the creation of large recycling organizations that may be profitable, either by law or economies of scale, the poor are more likely to be driven out of the recycling and the re-manufacturing market. To compensate for this loss of income, a society may need to create additional forms of societal programs to help support the poor. Like the parable of the broken window, there is a net loss to the poor and possibly the whole of a society to make recycling artificially profitable e.g. through the law. However, in Brazil and Argentina, waste pickers/informal recyclers work alongside the authorities, in fully or semi-funded cooperatives, allowing informal recycling to be legitimized as a paid public sector job.

Because the social support of a country is likely to be less than the loss of income to the poor undertaking recycling, there is a greater chance the poor will come in conflict with the large recycling organizations. This means fewer people can decide if certain waste is more economically reusable in its current form rather than being reprocessed. Contrasted to the recycling poor, the efficiency of their recycling may actually be higher for some materials because individuals have greater control over what is considered "waste."

One labor-intensive underused waste is electronic and computer waste. Because this waste may still be functional and wanted mostly by those on lower incomes, who may sell or use it at a greater efficiency than large recyclers.

Some recycling advocates believe that laissez-faire individual-based recycling does not cover all of society's recycling needs. Thus, it does not negate the need for an organized recycling program. Local government can consider the activities of the recycling poor as contributing to property blight.

Public Participation Rates

Changes that have been demonstrated to increase recycling rates include:

- Single-stream recycling

- Pay as you throw fees for trash

"Between 1960 and 2000, the world production of plastic resins increased 25-fold, while recovery of the material remained below 5 percent." Many studies have addressed recycling behaviour and strategies to encourage community involvement in recycling programmes. It has been argued that recycling behaviour is not natural because it requires a focus and appreciation for long-term planning, whereas humans have evolved to be sensitive to short-term survival goals; and that to overcome this innate predisposition, the best solution would be to use social pressure to compel participation in recycling programmes. However, recent studies have concluded that social pressure is unviable in this context. One reason for this is that social pressure functions well in small group sizes of 50 to 150 individuals (common to nomadic hunter–gatherer peoples) but not in communities numbering in the millions, as we see today. Another reason is that individual recycling does not take place in the public view.

In a study done by social psychologist Shawn Burn, it was found that personal contact with individuals within a neighborhood is the most effective way to increase recycling within a community. In his study, he had 10 block leaders talk to their neighbors and persuade them to recycle. A comparison group was sent fliers promoting recycling. It was found that the neighbors that were personally contacted by their block leaders recycled much more than the group without personal contact. As a result of this study, Shawn Burn believes that personal contact within a small group of people is an important factor in encouraging recycling. Another study done by Stuart Oskamp examines the effect of neighbors and friends on recycling. It was found in his studies that people who had friends and neighbors that recycled were much more likely to also recycle than those who didn't have friends and neighbors that recycled.

Many schools have created recycling awareness clubs in order to give young students an insight on recycling. These schools believe that the clubs actually encourage students to not only recycle at school, but at home as well.

Kerbside Collection

Kerbside collection, or curbside collection, is a service provided to households, typically in urban and suburban areas of removing household waste. It is usually accomplished by personnel using purpose built vehicles to pick up household waste in containers acceptable to or prescribed by the municipality.

Kerbside collection in Canberra, Australia

History

Prior to the 20th century the amount of waste generated by a household was relatively small. Household wastes were often simply thrown out the window, buried in the garden or deposited in outhouses. When human concentrations became more dense, waste collectors, called nightmen or gong farmers were hired to collect the night soil from pail closets, performing their duties only at night (hence the name). Meanwhile, disposing of refuse became a problem wherever cities grew. Often refuse was placed in unusable areas just outside the city, such as wetlands and tidal zones. One example is London, which from Roman times disposed of its refuse outside the London Wall beside the River Thames. Another example is 1830s Manhattan, where thousands of hogs were permitted to roam the streets and eat garbage. A small industry developed as "swill children" collected kitchen refuse to sell for pig feed and the rag and bone man traded goods for bones (used for glue) and rags (essential for paper manufacture prior to the invention of wood pulping). Later, in the late nineteenth century, trash was fed to swine in industrial.

As sanitation engineering came to be practised beginning in the mid-19th century and human waste was conveyed from the home in pipes, the gong farmer was replaced by the municipal trash collector as there remained growing amounts of household refuse, including fly ash from coal, which was burnt for home heating. In Paris, the rag and bone man worked side by side with the municipal bin man, though reluctantly: in 1884, Eugène Poubelle introduced the first integrated kerbside collection and recycling system, requiring residents to separate their waste into perishable items, paper and cloth, and crockery and shells. He also established rules for how private collectors and city workers should cooperate and he developed standard dimensions for refuse containers: his name in France is now synonymous with the garbage can. Under Poubelle, food waste and other organics collected in Paris were transported to nearby Saint Ouen where they were composted. This continued well into the 20th century when plastics began to contaminate the waste stream.

From the late-19th century to the mid-20th century, more or less consistent with the rise of consumables and disposable products municipalities began to pass anti-dumping ordinances and introduce kerbside collection. Residents were required to use a variety of refuse containers to facilitate kerbside collection but the main type was a variation of Poubelle's metal garbage container. It was not until the late 1960s that the green bin bag was introduced by Glad. Later, as waste management practices were introduced with the aim of reducing landfill impacts, a range of container types, mostly made of durable plastic, came to be introduced to facilitate the proper diversion of the waste stream. Such containers include blue boxes, green bins and wheelie bins or MGBs.

Over time, waste collection vehicles gradually increased in size from the hand pushed tip cart or English dust cart, a name by which these vehicles are still referred, to large compactor trucks.

Waste Management and Resource Recovery

Kerbside collection is today often referred to as a strategy of local authorities to collect recyclable items from the consumer. Kerbside collection is considered a low-risk strategy to reduce waste volumes and increase recycling rates. Materials are typically collected in large bins, coloured bags, or small open plastic tubs, specifically designated for content.

Glass for collection in Edinburgh, Scotland.

Recyclable materials that may be separately collected from municipal waste include:

Biodegradable waste component

- Green waste
- Kitchen waste

Recyclable materials, depending on location

- Office paper

- Newsprint

- Paperboard

- Corrugated fiberboard

- Plastics (#1 PET, #2 HDPE natural and colored, #3 PVC narrow-necked containers, #4 LDPE, #5 PP, #6 Polystyrene (however not EXPANDED polystyrene, an example of recyclable polystyrene may be a yoghurt pot) #7 other mixed resin plastics)

- Glass

- Copper

- Aluminum

- Steel and Tinplate

- Co-mingled recyclables- can be sorted by a clean materials recovery facility

In Somerville, MA all accepted paper, glass, plastic, and metal recycling is picked up from a single bin

Kerbside collection of recyclable resources is aimed to recover purer waste streams with higher market value than by other collection methods. If the household incorrectly separates the recyclable elements, the load may have to be put to landfill if it is deemed to be contaminated.

Kerbside collection and household recycling schemes are also being used as tools by local authorities to increase the public's awareness of their waste production.

Kerbside collection is commonly considered to be completely environmentally friendly. This may not necessarily be the case as it leads to an increased number of waste collection vehicles on the road, in themselves contributing to global warming through exhaust emissions until the time of their conversion to clean energy.

New and emerging waste treatment technologies such as mechanical biological treatment may offer an alternative to kerbside collection through automated separation of waste in recycling factories.

Usage by Country

Canada

Calgary, Alberta has adopted "Curbside" Recycling and uses blue bins. The blue cart programme accepts all types of recyclables, including plastics 1-7. It is picked up weekly for the cost of $8.00 per month. This programme is mandatory.

Halifax Regional Municipality (HRM) in Nova Scotia, Canada, with a population of about 375,000, has one of the most complex kerbside collection programmes in North America. Based on the green cart, it requires residents to self-sort refuse and place different types at the kerb on alternating weeks. As shown in the photo at left, week 1 would see the green cart and optional orange bags used for kitchen waste and other organics such as yard waste. Week 2 would permit non-recoverable waste in garbage bags or cans. Blue bags are used for paper, plastic and metal containers. Together with used grocery bags containing newspapers, they may be placed on the kerb either week. In summer, the green cart is emptied weekly due to the prevalence of flies. HRM has achieved a diversion rate of approximately 60 percent by this method.

In 1981 Resource Integration Systems (RIS) in collaboration with Laidlaw International tested the first blue box recycling system on 1500 homes in Kitchener, Ontario. Due to the success of the project the City of Kitchener put out a contract for public bid in 1984 for a recycling system city wide. Laidlaw won the bid and continued with the popular blue box recycling system. Today hundreds of cities around the world use the blue box system or a similar variation.

Many Canadian municipalities use "green bins" for kerbside recycling. Others, such as Moncton, use wet/dry waste separation and recovery programmes.

New Zealand

In New Zealand, kerbside collection of general refuse and recycling, and in some areas organic waste, is the responsibility of the local city or district council, or private contractors. Practices and collection methods vary widely from council to council and company to company. Some examples of collection are:

Kerbside collection bins in Dunedin, New Zealand. The yellow-liddied wheelie bin is for non-glass recyclables, and the blue bin is for glass. The two bins are collected on alternating weeks. Official council bags are used for general household waste, and are collected weekly.

- Auckland City Council: Two 240-litre wheelie bins are supplied: a red-lidded bin for general refuse, collected weekly, and a blue-lidded bin for recyclables, collected fortnightly.

- Christchurch City Council: Three wheelie bins are supplied: a 140-litre red-lidded bin for general refuse, a 240-litre yellow-lidded bin for recyclables, and an 80-litre green-lidded bin for organic waste. The organic waste bins are collected weekly, while the recyclables and general refuse bins are collected on alternating weeks.

- Hamilton City Council and Hutt City Council: A 45-litre bin is supplies for recyclables, collected weekly. General refuse is collected weekly using user-pays official council bags.

- Dunedin City Council, Palmerston North City Council and Wellington City Council: Two bins are supplied: a 45-litre or 70-litre bin for glass, and an 80-litre or 240-litre wheelie bin for non-glass recyclables. These two bins are collected on alternating weeks. General refuse is collected weekly using user-pays official council bags.

- Rodney District Council: A 45-litre bin is supplies for recyclables, collected weekly. There is no council collection of general waste, and all general waste collection is carried out by independent companies.

- Taupo District Council: A 45-litre bin is supplies for recyclables, collected weekly. General refuse is collected weekly using user-pays system of orange tags - one orange tag is to be placed on a standard rubbish bag up to 60 litres capacity, or half an orange sticker can be placed on two supermarket bags tied together.

- Upper Hutt City Council: Recycling is to be placed in plastic bags, with paper and cardboard collected in the first week, and plastic, metal and glass in the second week. General refuse is collected weekly using user-pays official council bags.

- Waitakere City Council: A 140-litre wheelie bin is provided for recyclables, collected fortnightly. General refuse is collected weekly using user-pays official council bags.

By 1996 the New Zealand cities of Auckland, Waitakere, North Shore and Lower Hutt had kerbside recycling bins available. In New Plymouth, Wanganui and Upper Hutt recyclable material was collected if placed in suitable bags. By 2007 73% of New Zealanders had access to kerbside recycling.

Kerbside collection of organic waste is carried out by the Mackenzie District Council and the Timaru District Council. Christchurch City Council is introducing the system to their kerbside collection. Other councils are carrying out trials.

United Kingdom

In the United Kingdom, the Household Waste Recycling Act 2003 requires local authorities to provide every household with a separate collection of at least two types of recyclable materials by 2010. There has been criticism in the difference of schemes used in the country such as the colour of bins, whether they are bins boxes or bags, and also the fact that clutter roads and how the additional trucks and collections needed have carbon dioxide emissions too. Some find the colour differences confusing, and people want a national scheme. A typical example is to compare two neighbouring councils in greater Manchester, Bury council and Salford. Bury uses blue for cans, plastic and glass, green for paper and cardboard and brown for garden waste. Salford uses blue for paper and card, brown for cans plastic and glass and pink for garden waste. Most councils use grey or black for general waste, with a few exceptions such as Liverpool, which uses purple for general waste, a colour used by no other council

Another controversial issue in the uk is the frequency of the waste collections. To save money, many councils are cutting the frequency of both general waste and recyclables collections. This has led to problems from larger families, and has led to overflowing and fly tipping. For example, previously, Bury Council collected general waste once a week and recyclables fortnightly. This has now changed to fortnightly for general waste and monthly (every 4 weeks) collection of recyclables.

A few councils are using "forced" recycling, by replacing the large, 240l general waste bin with a smaller 180l or 140l bin, and using the old 240l one for recyclables. This may be made worse by fortnightly collections of the "small" bin, and strict rules such as "No extra bags will be taken" and "Bin lids must be fully closed". Stockport Council is a notable user of this scheme. Their recycling rates have risen substantially as a result, but there are usually complaints from families. Trafford council also use a similar scheme, but the small grey bin is emptied every week. In addition, the two named councils, and more, collect food waste together with garden waste, by sending out kitchen caddies and compostable liners. These prevent food waste (including meat) from going to land-fill, and to increase the councils recycling rate. The food and garden waste is usually collected weekly or fortnightly, and is taken to an In Vessel composter or Anaerobic digester, where the waste is turned into soil improver for use on local farms.

In the north west, all the glass collected is used within the UK, around half of the plas-tics and cans are used in the UK; the rest is sent further afield to Europe or China to be made into new products, and paper and paperboard collected is sent to local paper mills to be made into newspapers, tissues, paperboard and office paper. Again some of the paper will be sent further afield.

Some councils only have 3 bins- general, organic and recyclables. This means that plas-tics, cans and glass go in the same container as paper and cardboard. Although this is much easier for the residents, there is more sorting required, and the paper quality is sometimes of a low grade due to food contamination or shards of glass in the paper, and so this scheme is criticized.

Also, most councils require residents to remove caps from bottles and rinse them out to avoid smells. This is because the lids are made from a different type of plastic (PP) to the bottle (PET/HDPE) - although by collapsing the bottles and folding them over like toothpaste tubes and rescrewing the caps in place enables the volume of bottles to be drastically reduced, thereby increasing the amount of bottles that can be carried in the recycling bins. In fact many bottlers, especially bottled water companies, have now designed their bottles to be collapsible; though this message has not been effectively disseminated to the consumer. A collapsable bottle takes between 25% and 33% of the space a non-collapsed bottle.

Labels are rarely required to be removed, however. This also means that only plastic bottles are recycled. Councils are still trying to make clear that plastic tubs (yogurts, desserts and spreads), bags and cling film cannot be recycled through the kerbside eco-nomically. If too much contamination is collected then this results in the whole vehicle load going to landfill at a high cost. Contamination is usually a problem if recyclables are collected in wheelie bins, as the worker can only look at the top; there may be con-tamination 'hidden' at the bottom. Councils that use many bags and boxes (Edinburgh) suffer from less contamination but are complicated and the loose paper and cardboard, and recycling bags are blown around, and paper can be wet.

Basque Country

In the province of Gipuzkoa, this system is implanted in many towns as Usurbil, Hernani, Oiartzun, Antzuola, Legorreta, Itsasondo, Zaldibia, Anoeta, Alegia, Irura,

Zizurkil, Astigarraga, Ordizia, Oñati and Lezo, where the common used name in basque is atez-atekoa, which means *door-by-door*. Due to the big success in this towns, with more than the 80% of the waste recycled, 34 towns in Gipuzkoa are studying to set this system up in 2013, like Arrasate, Bergara, Aretxabaleta, Eskoriatza, Legazpi, Tolosa or Pasaia.

The atez-ate system consists in hanging each kind of rubbish in a hanger outside the house a certain day or days in a week. For example, in Hernani, they have three days to hang their organic rubbish, two days for plastics and metallics, one for paper and one for rejects residuals.

This system started in the town of Usurbil in the year 2009, due to the incinerator of the region of Gipuzkoa was going to build in this town, exactly in the neighborhood of Zubieta. Three years after, the construction of the incinerator was paralyzed by the government of the region, suggesting that the incinerator was a source of contamination and the high cost of the building.

Criticism

This type of collection service is subject to growing criticism.

- The large (Wheelie bin) container encourages the "out of sight" rubbish mentality and invites more rubbish to be disposed.

- The bins and collection trucks are not suited to narrow roads or houses with steep driveways or steps.

- They lock local authorities into capital intensive equipment programmes and multi-national providers.

- Co-mingled recyclables are sometimes not being successfully managed by automated sorting stations and the rates of diversion are low. In some cases, this results in mountains of unsorted recyclables.

- In the UK especially, some councils are sending out at least 4 large bins - residents of smaller houses with no gardens have little space to put them

- Many use small plastic boxes, bags and lockable outdoor food waste 'caddies' which get blown around and lost, bad for recycling participation.

Computer Recycling

Computer recycling, electronic recycling or e-waste recycling is the disassembly and separation of components and raw materials of waste electronics. Although the procedures of re-use, donation and repair are not strictly recycling, they are other common sustainable ways to dispose of IT waste.

Computer monitors are typically packed into low stacks on wooden pallets for recycling and then shrink-wrapped.

In 2009, 38% of computers and a quarter of total electronic waste was recycled in the United States, 5% and 3% up from 3 years prior respectively. Since its inception in the early 1990s, more and more devices are recycled worldwide due to increased awareness and investment. Electronic recycling occurs primarily in order to recover valuable rare earth metals and precious metals, which are in short supply, as well as plastics and metals. These are resold or used in new devices after purification, in effect creating a circular economy.

Recycling is considered environmentally friendly because it prevents hazardous waste, including heavy metals and carcinogens, from entering the atmosphere, landfill or waterways. While electronics consist a small fraction of total waste generated, they are far more dangerous. There is stringent legislation designed to enforce and encourage the sustainable disposal of appliances, the most notable being the Waste Electrical and Electronic Equipment Directive of the European Union and the United States National Computer Recycling Act.

Opponents argue that recycling is expensive and ineffective, that it does not safeguard data and that it stifles innovation. It is also criticised for exporting, often illegally, large volumes of toxic waste to countries such as India, China and Nigeria for crude manual disassembly by workers who have little regard for the risk to themselves or the environment.

Reasons for Recycling

Obsolete computers and old electronics are valuable sources for secondary raw materials if recycled; otherwise, these devices are a source of toxins and carcinogens. Rapid technology change, low initial cost, and planned obsolescence have resulted in a fast-growing surplus of computers and other electronic components around the globe. Technical solutions are available, but in most cases a legal framework, collection system, logistics, and other services need to be implemented before applying a technical solution. The U.S. Environmental Protection Agency, estimates 30 to 40 million surplus PCs, classified as "hazardous household waste", would be ready for end-of-life management in the next few years. The U.S. National Safety Council estimates that 75% of all personal computers ever sold are now surplus electronics.

In 2007, the United States Environmental Protection Agency (EPA) stated that more than 63 million computers in the U.S. were traded in for replacements or discarded. Today, 15% of electronic devices and equipment are recycled in the United States. Most electronic waste is sent to landfills or incinerated, which releases materials such as lead, mercury, or cadmium into the soil, groundwater, and atmosphere, thus having a negative impact on the environment.

Many materials used in computer hardware can be recovered by recycling for use in future production. Reuse of tin, silicon, iron, aluminium, and a variety of plastics that are present in bulk in computers or other electronics can reduce the costs of constructing new systems. Components frequently contain lead, copper, gold and other valuable materials suitable for reclamation.

Computer components contain many toxic substances, like dioxins, polychlorinated biphenyls (PCBs), cadmium, chromium, radioactive isotopes and mercury. A typical computer monitor may contain more than 6% lead by weight, much of which is in the lead glass of the cathode ray tube (CRT). A typical 15 inch (38 cm) computer monitor may contain 1.5 pounds (1 kg) of lead but other monitors have been estimated to have up to 8 pounds (4 kg) of lead. Circuit boards contain considerable quantities of lead-tin solders that are more likely to leach into groundwater or create air pollution due to incineration. The processing (e.g. incineration and acid treatments) required to reclaim these precious substances may release, generate, or synthesize toxic byproducts.

Export of waste to countries with lower environmental standards is a major concern. The Basel Convention includes hazardous wastes such as, but not limited to, CRT screens as an item that may not be exported transcontinentally without prior consent of both the country exporting and receiving the waste. Companies may find it cost-effective in the short term to sell outdated computers to less developed countries with lax regulations. It is commonly believed that a majority of surplus laptops are routed to developing nations as "dumping grounds for e-waste". The high value of working and reusable laptops, computers, and components (e.g. RAM) can help pay the cost of transportation for many worthless "commodities". The laws governing the exportation of waste electronics are put in place to stop "recycling companies" in developed countries from shipping their waste to 3rd world countries as working devices; they are never working devices. The 3rd world workers scavenge specific items with selling value and throw the rest away to rot and become a health hazard in their own backyard.

Recycling one computer and monitor can save 539 pounds of fossil fuel, 48 pounds of chemicals, and 1.5 tons of water .

Regulations

An abandoned Taxan monitor.

Europe

In Switzerland, the first electronic waste recycling system was implemented in 1991, beginning with collection of old refrigerators; over the years, all other electric and electronic devices were gradually added to the system. The established producer responsibility organization is SWICO, mainly handling information, communication, and organization technology. The European Union implemented a similar system in February 2003, under the Waste Electrical and Electronic Equipment Directive (WEEE Directive, 2002/96/EC).

Pan European adoption of the Legislation was slow on take-up, with Italy and the United Kingdom being the final member states to pass it into law. The success of the WEEE directive has varied significantly from state to state, with collection rates varying between 13 kilograms per capita per annum to as little as 1 kg per capita per annum. Computers & electronic wastes collected from households within Europe are treated under the WEEE directive via Producer Compliance Schemes (whereby manufacturers of Electronics pay into a scheme that funds its recovery from household waste recycling centres (HWRCs)) and nominated Waste Treatment Facilities (known as Obligated WEEE).

However, recycling of ex corporate Computer Hardware and associated electronic equipment falls outside the Producer Compliance Scheme (Known as non-obligated). In the UK, Waste or obsolete corporate related computer hardware is treated via third party Authorized Treatment Facilities, who normally impose a charge for its collection and treatment.

United States

Federal

The United States Congress considers a number of electronic waste bills, like the National Computer Recycling Act introduced by Congressman Mike Thompson (D-CA).

The main federal law governing solid waste is the Resource Conservation and Recovery Act of 1976. It covers only CRTs, though state regulations may differ. There are also separate laws concerning battery disposal. On March 25, 2009, the House Science and Technology Committee approved funding for research on reducing electronic waste and mitigating environmental impact, regarded by sponsor Ralph Hall (R-TX) as the first federal bill to directly address electronic waste.

State

Many states have introduced legislation concerning recycling and reuse of computers or computer parts or other electronics. Most American computer recycling legislations address it from within the larger electronic waste issue.

In 2001, Arkansas enacted the Arkansas Computer and Electronic Solid Waste Management Act, which requires that state agencies manage and sell surplus computer equipment, establishes a computer and electronics recycling fund, and authorizes the Department of Environmental Quality to regulate and/or ban the disposal of computer and electronic equipment in Arkansas landfills.

The recently passed Electronic Device Recycling Research and Development Act distributes grants to universities, government labs and private industries for research in developing projects in line with e-waste recycling and refurbishment.

Asia

In Japan, sellers and manufacturers of certain electronics (such as televisions and air conditioners) are required to recycle them. However, no legislation exists to cover the recycling of computer or cellphone related wastes.

It is required in South Korea and Taiwan that sellers and manufacturers of electronics be responsible for recycling 75% of their used products.

According to a report by UNEP titled, "Recycling - from E-Waste to Resources," the amount of e-waste being produced - including mobile phones and computers - could rise by as much as 500 percent over the next decade in some countries, such as India.

Electronic waste is often exported to developing countries.

One theory is that increased regulation of electronic waste and concern over the environmental harm in mature economies creates an economic disincentive to remove residues prior to export. Critics of trade in used electronics maintain that it is too easy for brokers calling themselves recyclers to export unscreened electronic waste to developing countries, such as China, India and parts of Africa, thus avoiding the expense of removing items like bad cathode ray tubes (the processing of which is expensive and difficult). The developing countries are becoming big dump yards of e-waste. Proponents of international trade point to the success of fair trade programs in other industries, where cooperation has led creation of sustainable jobs, and can bring affordable technology in countries where repair and reuse rates are higher.

4.5-volt, D, C, AA, AAA, AAAA, A23, 9-volt, CR2032 and LR44 cells are all recyclable in most countries.

Organizations like A2Z Group (Company Website) have stepped in to own up the responsibility to collect and recycle e-waste at various locations in India.

South Africa

Thanks to the National Environmental Management Act 1998 and National Environmental Management Waste Act 2008, any person in any position causing harm to the environment and failing to comply with the Waste Act could be fined R10 Million or put into jail or receive both penalties for their transgressions.

Recycling Methods

Computers being collected for recycling at a pickup event in Olympia, Washington, United States.

Consumer Recycling

Consumer recycling options consists of sale, donating computers directly to organizations in need, sending devices directly back to their original manufacturers, or getting components to a convenient recycler or refurbisher.

Scrapping/Recycling

The rising price of precious metals — coupled with the high rate of unemployment during the Great Recession — has led to a larger number of amateur "for profit" electronics recyclers. Computer parts, for example, are stripped of their most valuable components and sold for scrap. Metals like copper, aluminum, lead, gold and palladium are recovered from computers, televisions and more.

In the recycling process, TVs, monitors, mobile phones and computers are typically tested for reuse and repaired. If broken, they may be disassembled for parts still having high value if labour is cheap enough. Other e-waste is shredded to roughly 100 mm pieces and manually checked to separate out toxic batteries and capacitors which contain poisonous metals. The remaining pieces are further shredded to ~10 mm and passed under a magnet to remove ferrous metals. An eddy current ejects non-ferrous metals, which are sorted by density either by a centrifuge or vibrating plates. Precious metals can be dissolved in acid, sorted, and smelted into ingots. The remaining glass and plastic fractions are separated by density and sold to re-processors. TVs and monitors must be manually disassembled to remove either toxic lead in CRTs or the mercury in flat screens.

Bulk laptops at a recycling affiliate, broken down into Dell, Gateway Computers, Hewlett-Packard, Sony, and other.

Corporate Recycling

Businesses seeking a cost-effective way to recycle large amounts of computer equipment responsibly face a more complicated process.

Businesses also have the options of sale or contacting the Original Equipment Manufacturers (OEMs) and arranging recycling options.

Some companies pick up unwanted equipment from businesses, wipe the data clean from the systems, and provide an estimate of the product's remaining value. For unwanted items that still have value, these firms buy the excess IT hardware and sell refurbished products to those seeking more affordable options than buying new.

Companies that specialize in data protection and green disposal processes dispose of both data and used equipment, while employing strict procedures to help improve the environment. Professional IT Asset Disposition (ITAD) firms specialize in corporate computer disposal and recycling services in compliance with local laws and regulations and also offer secure data elimination services that comply with Data remanence standards including National Institute of Standards and Technology.

Corporations face risks both for incompletely destroyed data and for improperly disposed computers. In the UK, some recycling companies use a specialized WEEE-registered contractor to dispose IT equipment and electrical appliances, who disposes it safely and legally. In America, companies are liable for compliance with regulations even if the recycling process is outsourced under the Resource Conservation and Recovery Act. Companies can mitigate these risks by requiring waivers of liability, audit trails, certificates of data destruction, signed confidentiality agreements, and random audits of information security. The National Association of Information Destruction is an international trade association for data destruction providers.

Sale

Online auctions are an alternative for consumers willing to resell for cash less fees, in a complicated, self-managed, competitive environment where paid listings might not sell. Online classified ads can be similarly risky due to forgery scams and uncertainty.

Take Back

When researching computer companies before a computer purchase, consumers can find out if they offer recycling services. Most major computer manufacturers offer some form of recycling. At the user's request they may mail in their old computers, or arrange for pickup from the manufacturer.

Hewlett-Packard also offers free recycling, but only one of its "national" recycling programs is available nationally, rather than in one or two specific states. Hewlett-Packard also offers to pick up any computer product of any brand for a fee, and to offer a coupon against the purchase of future computers or components; it was the largest computer recycler in America in 2003, and it has recycled over 750,000,000 pounds (340,000,000 kg) of electronic waste globally since 1995. It encourages the shared approach of collection points for consumers and recyclers to meet.

Exchange

Manufacturers often offer a free replacement service when purchasing a new PC. Dell Computers and Apple Inc. take back old products when one buys a new one. Both refurbish and resell their own computers with a one-year warranty.

Many companies purchase and recycle all brands of working and broken laptops and notebook computers from individuals and corporations. Building a market for recycling of desktop computers has proven more difficult than exchange programs for laptops, smartphones and other smaller electronics. A basic business model is to provide a seller an instant online quote based on laptop characteristics, then to send a shipping label and prepaid box to the seller, to erase, reformat, and process the laptop, and to pay rapidly by cheque. A majority of these companies are also generalized electronic waste recyclers as well; organizations that recycle computers exclusively include Cash For Laptops, a laptop refurbisher in Nevada that claims to be the first to buy laptops online, in 2001.

Donations/Nonprofits

With the constant rising costs due to inflation, many families or schools do not have the sufficient funds available for computers to be utilized along with education standards. Families also impacted by disaster suffer as well due to the financial impact of the situation they have incurred. Many nonprofit organizations, such as InterConnection.org, can be found locally as well as around the web and give detailed descriptions as to what methods are used for dissemination and detailed instructions on how to donate. The impact can be seen locally and globally, affecting thousands of those in need. In Canada non profit organizations engaged in computer recycling, such as The Electronic Recycling Association Calgary, Edmonton, Vancouver, Winnipeg, Toronto, Montreal, Computers for Schools Canada wide, are very active in collecting and refurbishing computers and laptops to help the non profit and charitable sectors and schools.

Junkyard Computing

The term *junkyard computing* is a colloquial expression for using old or inferior hardware to fulfill computational tasks while handling reliability and availability on software level. It utilizes abstraction of computational resources via software, allowing hardware replacement at very low effort. Ease of replacement is hereby a corner point since hardware failures are expected at any time due to the condition of the underlying infrastructure. This paradigm became more widely used with the introduction of cluster orchestration software like Kubernetes or Apache Mesos, since large monolithic applications require reliability and availability on machine level whereas this kind of software is fault tolerant by design. Those orchestration tools also introduced fairly fast set-up processes allowing to use junkyard computing economically and even making this pattern applicable in the first place. Further use cases were introduced when continuous delivery was getting more widely accepted. Infrastructure to execute tests and static code analysis was needed which requires as much performance as possible while being extremely cost effective. From an economical and technological perspective, junkyard computing is only practicable for a small amount of users or companies. It already requires a descend amount of physical machines to compensate hardware failures while maintaining the required reliability and availability. This implies a direct need for a matching underling infrastructure to house all the computers and servers. Scaling this paradigm is also quiet limited due to the increasing importance of factors like power efficiency and maintenance efforts, making this kind of computing perfect for mid-sized applications.

History

Although consumer electronics such as the radio have been popular since the 1920s, recycling was almost unheard of until the early 1990s. At the end of the 1970s the accelerating pace of domestic consumer electronics drastically shortened the lifespan of electronics such as TVs, VCRs and audio. New innovations appeared more quickly, making older equipment considered obsolete. Increased complexity and sophistication of manufacture made local repair more difficult. The retail market shifted gradually, but substantially from a few high-value items that were cherished for years and repaired when necessary, to short-lived items that were rapidly replaced owing to wear or simply fashion, and discarded rather than repaired. This was particularly evident in computing, highlighted by Moore's Law. In 1988 two severe incidents highlighted the approaching e-waste crisis. The cargo barge Khian Sea, was loaded with more than 14,000 tons of toxic ash from Pennsylvania which had been refused acceptance in New Jersey and the Caribbean. After sailing for 16 months, all the waste was dumped as "topsoil fertiliser" in Haiti and in the Bay of Bengal by November 1988. In June 1988, a large illegal toxic waste dump which had been created by an Italian company was discovered. This led to the formation of the Basel Convention to stem the flow of poisonous substances from developed countries in 1989.

In 1991, the first electronic waste recycling system was implemented in Switzerland, beginning with collection of old refrigerators but gradually expanding to cover all devices. The organisation SWICO handles the programme, and is a partnership between IT retailers.

The first publication to report the recycling of computers and electronic waste was published on the front page of the New York Times on April 14, 1993 by columnist Steve Lohr. It detailed the work of Advanced Recovery Inc., a small recycler, in trying to safely dismantle computers, even if most waste was landfilled. Several other companies emerged in the early 1990s, chiefly in Europe, where national 'take back' laws compelled retailers to use them.

After these schemes were set up, many countries did not have the capacity to deal with the sheer quantity of e-waste they generated or its hazardous nature. They began to export the problem to developing countries without enforced environmental legislation. This is cheaper: the cost of recycling of computer monitors in the US is ten times more than in China. Demand in Asia for electronic waste began to grow when scrap yards found they could extract valuable substances such as copper, iron, silicon, nickel and gold, during the recycling process.

The Waste Electrical and Electronic Equipment Directive (WEEE Directive) became European Law in February 2003 and covers all aspects of recycling all types of appliance. This was followed by Electronic Waste Recycling Act, enshrined in Californian law in January 2005

The 2000s saw a large increase in both the sale of electronic devices and their growth as a waste stream: in 2002 e-waste grew faster than any other type of waste in the EU. This caused investment in modern, automated facilities to cope with the influx of redundant appliances.

E-cycling

"E-cycling" or "E-waste" is an initiative by the United States Environmental Protection Agency (EPA) which refers to donations, reuse, shredding and general collection of used electronics. Generically, the term refers to the process of collecting, brokering, disassembling, repairing and recycling the components or metals contained in used or discarded electronic equipment, otherwise known as electronic waste (e-waste). "E-cyclable" items include, but are not limited to: televisions, computers, microwave ovens, vacuum cleaners, telephones and cellular phones, stereos, and VCRs and DVDs just about anything that has a cord, light or takes some kind of battery.

Investment in e-cycling facilities has been increasing recently due to technology's rapid rate of obsolescence, concern over improper methods, and opportunities for manufacturers to influence the secondary market (used and reused products). The higher metal prices is also having more recycling taking place. The controversy around methods stems from a lack of agreement over preferred outcomes.

World markets with lower disposable incomes, consider 75% repair and reuse to be valuable enough to justify 25% disposal. Debate and certification standards may be leading to better definitions, though civil law contracts, governing the expected process are still vital to any contracted process, as poorly defined as "e-cycling".

Pros of E-cycling

The e-waste disposal occurring after processing for reuse, repair of equipment, and recovery of metals may be unethical or illegal when e-scrap of many kinds is transported overseas to developing countries for such processing. It is transported as if to be repaired and/ or recycled, but after processing the less valuable e-scrap becomes e-waste/pollution there. Another point of view is that the net environmental cost must be compared to and include the mining, refining and extraction with its waste and pollution cost of new products manufactured to replace secondary products which are routinely destroyed in wealthier nations, and which cannot economically be repaired in older or obsolete products. As an example of negative impacts of e-waste, pollution of groundwater has become so serious in areas surrounding China's landfills that water must be shipped in from 18 miles (29 km) away. However, mining of new metals can have even broader impacts on groundwater. Either thorough e-cycling processing, domestic processing or overseas repair, can help the environment by avoiding pollution. Such e-cycling can theoretically be a sustainable alternative to disposing of e-waste in landfills. In addition, e-cycling allows for the reclamation of potential conflict minerals, like gold and wolframite, which requires less of those to be mined and lessens the potential money flow to militias and other exploitative actors in third-world that profit from mining them.

Supporters of one form of "required e-cycling" legislation argue that e-cycling saves taxpayers money, as the financial responsibility would be shifted from the taxpayer to the manufacturers. Advocates of more simple legislation (such as landfill bans for e-waste) argue that involving manufacturers does not reduce the cost to consumers, if reuse value is lost, and the resulting costs are then passed on to consumers in new products, particularly affecting markets which can hardly afford new products. It is theorized that manufacturers who take part in e-cycling would be motivated to use fewer materials in the production process, create longer lasting products, and implement safer, more efficient recycling systems. This theory is sharply disputed and has never been demonstrated.

Criticisms of E-cycling

The critics of e-cycling are just as vocal as its advocates. According to the Reason Foundation, e-cycling only raises the product and waste management costs of e-waste for consumers and limits innovation on the part of high-tech companies. They also believe that e-cycling facilities could unintentionally cause great harm to the environment. Critics claim that e-waste doesn't occupy a significant portion of total waste. According to a European study, only 4% of waste is electronic.

Another opposition to e-cycling is that many problems are posed in disassembly: the process is costly and dangerous because of the heavy metals of which the electronic products are composed, and as little as 1-5% of the original cost of materials can be retrieved. A final problem that people find is that identity fraud is all too common in regards to the disposal of electronic products. As the programs are legislated, creating winners and losers among e-cyclers with different locations and processes, it may be difficult to distinguish between criticism of e-cycling as a practice, and criticism of the specific legislated means proposed to enhance it.

The Fate of E-waste

A hefty criticism often lobbed at reuse based recyclers is that people think that they are recycling their electronic waste, when in reality it is actually being exported to developing countries like China, India, and Nigeria. For instance, at free recycling drives, "recyclers" may not be staying true to their word, but selling e-waste overseas or to parts brokers. Studies indicate that 50-80% of the 300,000 to 400,000 tons (270,000 to 360,000 tonnes) of e-waste is being sent overseas, and that approximately 2 million tons (1.8 million tonnes) per year go to U.S. landfills.

Although not possible in all circumstances, the best way to e-cycle is to upcycle e-waste. On the other hand, the electronic products in question are generally manufactured, and repaired under warranty, in the same nations, which anti-reuse recyclers depict as primitive. Reuse-based e-recyclers believe that fair-trade incentives for export markets will lead to better results than domestic shredding. There has been a continued debate between export-friendly e-cycling and increased regulation of that practice.

In the European Union, debate regarding the export of e-waste has resulted in a significant amendment to the WEEE directive (January 2012) with a view to significantly diminishing the export of WEEE (untreated e-waste). During debate in Strasburg, MEPs stated that "53 million tonnes of WEEE were generated in 2009 but only 18% collected for recycling" with the remainder being exported or sent to landfill. The Amendment, voted through by a unanimous 95% of representatives, removed the re-use (repair and refurbishmet) aspect of the directive and placed more emphasis upon recycling and recovery of precious metals and base metals. The changes went further by placing the burden upon registered exporters to prove that used equipment leaving Europe was "fit for purpose".

Policy Issues and Current Efforts

Currently, pieces of government legislation and a number of grassroots efforts have contributed to the growth of e-cycling processes which emphasize decreased exports over increased reuse rates. The Electronic Waste Recycling Act was passed in California in 2003. It requires that consumers pay an extra fee for certain types of electronics, and the collected money be then redistributed to recycling companies that are qualified to properly recycle these products. It is the only state that legislates against e-waste through this kind of consumer fee; the other states' efforts focus on producer responsibility laws or waste disposal bans. No study has shown that per capita recovery is greater in one type of legislated program (e.g. California) versus ordinary waste disposal bans (e.g. Massachusetts), though recovery has greatly increased in states which use either method.

As of September, 2006, Dell developed the nation's first completely free recycling program, furthering the responsibilities that manufacturers are taking for e-cycling. Manufacturers and retailers such as Best Buy, Sony, and Samsung have also set up recycling programs. This program does not accept televisions, which are the most expensive used electronic item, and are unpopular in markets which must deal with televisions when the more valuable computers have been cherry picked.

Another step being taken is the recyclers' pledge of true stewardship, sponsored by the Computer TakeBack Campaign. It has been signed by numerous recyclers promising to recycle responsibly. Grassroots efforts have also played a big part in this issue, as they and other community organizations are being formed to help responsibly recycle e-waste. Other grassroots campaigns are Basel, the Computer TakeBack Campaign (co-coordinated by the Grassroots Recycling Network), and the Silicon Valley Toxics Coalition. No study has shown any difference in recycling methods under the Pledge, and no data is available to demonstrate difference in management between "Pledge" and non-Pledge companies, though it is assumed that the risk of making false claims will prevent Pledge companies from wrongly describing their processes.

Many people believe that the U.S. should follow the European Union model in regards to its management of e-waste. In this program, a directive forces manufacturers to take responsibility for e-cycling; it also demands manufacturers' mandatory take-back and places bans on exporting e-waste to developing countries. Another longer-term solution is for computers to be composed of less dangerous products and many people disagree. No data has been provided to show that people who agree with the European model have based their agreement on measured outcomes or experience-based scientific method.

Data Security

E-waste presents a potential security threat to individuals and exporting countries. Hard drives that are not properly erased before the computer is disposed of can be reopened, exposing sensitive information. Credit card numbers, private financial data, account information and records of online transactions can be accessed by most willing individuals. Organized criminals in Ghana commonly search the drives for information to use in local scams.

Electronic waste dump at Agbogbloshie, Ghana. Organized criminals commonly search the drives for information to use in local scams.

Government contracts have been discovered on hard drives found in Agbogbloshie, Ghana. Multimillion-dollar agreements from United States security institutions such as the Defense Intelligence Agency (DIA), the Transportation Security Administration and Homeland Security have all resurfaced in Agbogbloshie.

Reasons to Destroy and Recycle Securely

There are ways to ensure that not only hardware is destroyed but also the private data on the hard drive. Having customer data stolen, lost, or misplaced contributes to the ever growing number of people who are affected by identity theft, which can cause corporations to lose more than just money. The image of a company that holds secure data, such as banks, law firms, pharmaceuticals, and credit corporations is also at risk. If a company's public image is hurt, it could cause consumers to not use their services and could cost millions in business losses and positive public relation campaigns. The cost of data breaches "varies widely, ranging from $90 to $50,000 (under HIPAA's new HITECH amendment, that came about through the American Recovery and Revitalization act of 2009), as per customer record, depending on whether the breach is "low-profile" or "high-profile" and the company is in a non-regulated or highly regulated area, such as banking or medical institutions."

There is also a major backlash from the consumer if there is a data breach in a company that is supposed to be trusted to protect their private information. If an organization has any consumer info on file, they must by law (Red Flags Clarification act of 2010) have written information protection policies and procedures in place, that serve to combat, mitigate, and detect vulnerable areas that could result in identity theft. The United States Department of Defense has published a standard to which recyclers and individuals may meet in order to satisfy HIPAA requirements.

Secure Recycling

Countries have developed standards, aimed at businesses and with the purpose of ensuring the security of Data contained in 'confidential' computer media [NIST 800-88: US standard for Data Remenance][HMG CESG IS5, Baseline & Enhanced, UK Government Protocol for Data Destruction]. National Association for Information Destruction (NAID) "is the international trade association for companies providing information destruction services. Suppliers of products, equipment and services to destruction companies are also eligible for membership. NAID's mission is to promote the information destruction industry and the standards and ethics of its member companies." There are companies that follow the guidelines from NAID and also meet all Federal EPA and local DEP regulations.

The typical process for computer recycling aims to securely destroy hard drives while still recycling the byproduct. A typical process for effective computer recycling:

1. Receive hardware for destruction in locked and securely transported vehicles.

2. Shred hard drives.

3. Separate all aluminum from the waste metals with an electromagnet.

4. Collect and securely deliver the shredded remains to an aluminum recycling plant.

5. Mold the remaining hard drive parts into aluminum ingots.

The Asset Disposal and Information Security Alliance (ADISA) publishes an *ADISA IT Asset Disposal Security Standard* that covers all phases of the e-waste disposal process from collection to transportation, storage and sanitization's at the disposal facility. It also conducts periodic audits of disposal vendors.

Plastic Recycling

Plastic recycling is the process of recovering scrap or waste plastic and reprocessing the material into useful products. Since plastic is non-biodegradable, recycling it is a part of global efforts to reduce plastic in the waste stream, especially the approximately eight million metric tonnes of waste plastic that enter the Earth's ocean every year. This helps to reduce the high rates of plastic pollution.

Plastic recycling includes taking any type of plastic sorting it into different polymers and then chipping it and then melting it down into pellets after this stage it can then be used to make items of any kind such as plastic chairs and tables. Soft Plastics are also recycled such as polyethylene film and bags. This closed-loop operation has taken place since the 1970s and has made the production of some plastic products amongst the most efficient operations today.

Compared with lucrative recycling of metal, and similar to the low value of glass, plastic polymers recycling is often more challenging because of low density and low value. There are also numerous technical hurdles to overcome when recycling plastic.

A macro molecule interacts with its environment along its entire length, so total energy involved in mixing it is largely due to the product side stoichiometry. Heating alone is not enough to dissolve such a large molecule, so plastics must often be of nearly identical composition to mix efficiently.

When different types of plastics are melted together, they tend to phase-separate, like oil and water, and set in these layers. The phase boundaries cause structural weakness in the resulting material, meaning that polymer blends are useful in only limited applications.

Another barrier to recycling is the widespread use of dyes, fillers, and other additives in plastics. The polymer is generally too viscous to economically remove fillers, and would be damaged by many of the processes that could cheaply remove the added dyes.

Additives are less widely used in beverage containers and plastic bags, allowing them to be recycled more often. Yet another barrier to removing large quantities of plastic from the waste stream and landfills is the fact that many common but small plastic items lack the universal triangle recycling symbol and accompanying number. An example is the billions of plastic utensils commonly distributed at fast food restaurants or sold for use at picnics.

The percentage of plastic that can be fully recycled, rather than downcycled or go to waste can be increased when manufacturers of packaged goods minimize mixing of packaging materials and eliminate contaminants. The Association of Plastics Recyclers have issued a Design Guide for Recyclability.

The use of biodegradable plastics is increasing.

Processes

Before recycling, most plastics are sorted according to their resin type. In the past, plastic reclaimers used the resin identification code (RIC), a method of categorization of polymer types, which was developed by the Society of the Plastics Industry in 1988. polyethylene terephthalate, commonly referred to as PET, for instance, has a resin code of 1. Most plastic reclaimers do not rely on the RIC now; they use automatic sort systems to identify the resin. Ranging from manual sorting and picking of plastic materials; to mechanized automation processes that involve shredding, sieving, separation by rates of density i.e. air, liquid, or magnetic, and complex spectrophotometric distribution technologies e.g. UV/VIS, NIR, Laser, etc. Some plastic products are also separated by color before they are recycled. The plastic recyclables are then shredded. These shredded fragments then undergo processes to eliminate impurities like paper labels. This material is melted and often extruded into the form of pellets which are then used to manufacture other products.

Thermal Depolymerization

Another process involves the conversion of assorted polymers into petroleum by a much less precise thermal depolymerization process. Such a process would be able to accept almost any polymer or mix of polymers, including thermoset materials such as vulcanized rubber tire separation of wastes and the biopolymers in feathers and other agricultural waste. Like natural petroleum, the chemicals produced can be made into fuels as well as polymers. A pilot plant of this type exists in Carthage, Missouri, United States, using turkey waste as input material. Gasification is a similar process, but is not technically recycling, since polymers are not likely to become the result.

Waste Plastic Pyrolysis to Fuel Oil

Plastic Pyrolysis can convert petroleum based waste streams such as plastics into quality fuels, carbons.

Given below is the list of suitable plastic raw materials for pyrolysis:

- Mixed plastic (HDPE, LDPE, PE, PP, Nylon, Teflon, PS, ABS, FRP etc.)

- Mixed waste plastic from waste paper mill

- Multi Layered Plastic

Heat Compression

Yet another process that is gaining ground with startup companies (especially in Australia, United States and Japan) is heat compression. The heat compression process takes all unsorted, cleaned plastic in all forms, from soft plastic bags to hard industrial waste, and mixes the load in tumblers (large rotating drums resembling giant clothes dryers). The most obvious benefit to this method is the fact that all plastic is recyclable, not just matching forms. However, criticism rises from the energy costs of rotating the drums, and heating the post-melt pipes.

Distributed Recycling

For some waste plastics, recent technical devices called recyclebots enable a form of distributed recycling. Preliminary life-cycle analysis(LCA) indicates that such distributed recycling of HDPE to make filament of 3-D printers in rural regions is energetically favorable to either using virgin resin or conventional recycling processes because of reductions in transportation energy

Other Processes

A process has also been developed in which many kinds of plastic can be used as a carbon source in the recycling of scrap steel. There is also a possibility of mixed recycling of different plastics, which does not require their separation. It is called Compatibilization and requires use of special chemical bridging agents compatibilizers. It can help to keep the quality of recycled material and to skip often expensive and inefficient preliminary scanning of waste plastics streams and their separation/purification.

Applications

PET

Post-consumer polyethylene terephthalate (PET or PETE) containers are sorted into different colour fractions, and baled for onward sale. PET recyclers further sort the baled bottles and they are washed and flaked (or flaked and then washed). Non-PET fractions such as caps and labels are removed during this process. The clean flake is dried. Further treatment can take place e.g. melt filtering and pelletising or various treatments to produce food-contact-approved recycled PET (RPET).

RPET has been widely used to produce polyester fibres. This sorted post-consumer PET waste is crushed, chopped into flakes, pressed into bales, and offered for sale.

One use for this recycled PET that has recently started to become popular is to create fabrics to be used in the clothing industry. The fabrics are created by spinning the PET flakes into thread and yarn. This is done just as easily as creating polyester from brand new PET. The recycled PET thread or yarn can be used either alone or together with other fibers to create a very wide variety of fabrics. Traditionally these fabrics are used to create strong, durable, rough, products, such as jackets, coat, shoes, bags, hats, and accessories since they are usually too rough for direct skin contact and can cause irritation. However, these types of fabrics have become more popular as a result of the public's growing awareness of environmental issues. Numerous fabric and clothing manufacturers have capitalized on this trend.

Other major outlets for RPET are new containers (food-contact or non-food-contact) produced either by (injection stretch blow) moulding into bottles and jars or by thermoforming APET sheet to produce clam shells, blister packs and collation trays. These applications used 46% of all RPET produced in Europe in 2010. Other applications, such as strapping tape, injection-moulded engineering components and even building materials account for 13% of the 2010 RPET production.

In the United States the recycling rate for PET packaging was 31.2% in 2013, according to a report from The National Association for PET Container Resources (NAPCOR) and The Association of Postconsumer Plastic Recyclers (APR). A total of 1,798 million pounds was collected and 475 million pounds of recycled PET used out of a total of 5,764 million pounds of PET bottles.

HDPE

Plastic # 2, high-density polyethylene (HDPE) is a commonly recycled plastic. It is typically downcycled into plastic lumber, tables, roadside curbs, benches, truck cargo liners, trash receptacles, stationery (e.g. rulers) and other durable plastic products and is usually in demand.

PS

The resin identification code symbol for polystyrene

Most polystyrene products are currently not recycled due to the lack of incentive to invest in the compactors and logistical systems required. As a result, manufacturers cannot obtain sufficient scrap. Expanded polystyrene (EPS) scrap can easily be added to products such as EPS insulation sheets and other EPS materials for construction applications. When it is not used to make more EPS, foam scrap can be turned into clothes hangers, park benches, flower pots, toys, rulers, stapler bodies, seedling containers, picture frames, and architectural molding from recycled PS.

Recycled EPS is also used in many metal casting operations. Rastra is made from EPS that is combined with cement to be used as an insulating amendment in the making of concrete foundations and walls. Since 1993, American manufacturers have produced insulating concrete forms made with approximately 80% recycled EPS.

Other Plastics

The white plastic polystyrene foam peanuts used as packing material are often accepted by shipping stores for reuse.

Successful trials in Israel have shown that plastic films recovered from mixed municipal waste streams can be recycled into useful household products such as buckets.

Similarly, agricultural plastics such as mulch film, drip tape and silage bags are being diverted from the waste stream and successfully recycled into much larger products for industrial applications such as plastic composite railroad ties. Historically, these agricultural plastics have primarily been either landfilled or burned on-site in the fields of individual farms.

CNN reports that Dr. S. Madhu of the Kerala Highway Research Institute, India, has formulated a road surface that includes recycled plastic: aggregate, bitumen (asphalt) with plastic that has been shredded and melted at a temperature below 220 degrees C (428 °F) to avoid pollution. This road surface is claimed to be very durable and monsoon rain resistant. The plastic is sorted by hand, which is economical in India. The test road used 60 kg of plastic for an approximately 500m-long, 8m-wide, two-lane road. The process chops thin-film road-waste into a light fluff of tiny flakes that hot-mix plants can uniformly introduce into viscous bitumen with a customized dosing machine. Tests at both Bangalore and the Indian Road Research Centre indicate that roads built using this 'KK process' will have longer useful lives and better resistance to cold, heat, cracking, and rutting, by a factor of three.

Recycling Rates

The quantity of post-consumer plastics recycled has increased every year since at least 1990, but rates lag far behind those of other items, such as newspaper (about 80%) and corrugated fiberboard (about 70%). Overall, U.S. post-consumer plastic waste for 2008 was estimated at 33.6 million tons; 2.2 million tons (6.5%) were recycled and 2.6 million tons (7.7%) were burned for energy; 28.9 million tons, or 85.5%, were discarded in landfills.

Economic and Energy Potential

In 2008, the price of PET dropped from $370/ton in the US to $20 in November. PET prices had returned to their long-term averages by May 2009.

Recycling one ton of plastic can save 5,774 kWh of energy, 98,000,000 btus of energy, 1,000-2,000 gallons of gasoline, 685 gallons of oil, 30 cubic yards of landfill space, 48,000 gallons of water.

Consumer Education

United Kingdom

In the UK, the amount of post-consumer plastic being recycled is relatively low, due in part to a lack of recycling facilities.

The Plastics 2020 Challenge was founded in 2009 by the plastics industry with the aim of engaging the British public in a nationwide debate about the use, reuse and disposal of plastics, and hosts a series of online debates on its website framed around the waste hierarchy.

There is a facility in Worksop capable of recycling 60–80 thousand metric tonnes a year.

In Northern Ireland, the rate of recycling is relatively low at only 37.4%. However, emerging technologies are helping to increase the recycling rates of items previously landfilled e.g. mixed hard plastics.

Plastic Identification Code

Five groups of plastic polymers, each with specific properties, are used worldwide for packaging applications. Each group of plastic polymer can be identified by its Plastic Identification code (PIC), usually a number or a letter abbreviation. For instance, Low-Density Polyethylene can be identified by the number "4" or the letters "LDPE". The PIC appears inside a three-chasing-arrow recycling symbol. The symbol is used to indicate whether the plastic can be recycled into new products.

The PIC was introduced by the Society of the Plastics Industry, Inc., to provide a uniform system for the identification of various polymer types and to help recycling companies separate various plastics for reprocessing. Manufacturers of plastic products are required to use PIC labels in some countries/regions and can voluntarily mark their products with the PIC where there are no requirements. Consumers can identify the plastic types based on the codes usually found at the base or at the side of the plastic products, including food/chemical packaging and containers. The PIC is usually not present on packaging films, since it is not practical to collect and recycle most of this type of waste.

Plastic Identifica-tion Code	Type of plas-tic polymer	Properties	Common Packag-ing Applications	Melting-(°C) and Glass Tran-sition Tem-peratures	Young's Mod-ulus (GPa)
01 PET	Polyethylene terephthalate (PET, PETE)	Clarity, strength, toughness, barrier to gas and mois-ture.	Soft drink, water and salad dressing bottles; peanut butter and jam jars; small consumer electronics.	Tm = 250; Tg = 76	2-2.7
02 PE-HD	High-density polyethylene (HDPE)	Stiffness, strength, toughness, resistance to moisture, permeability to gas.	Water pipes, hula hoop rings, five gal-lon buckets, milk, juice and water bot-tles; grocery bags, some shampoo/toiletry bottles.	Tm = 130; Tg = -125	0.8
03 PVC	Polyvinyl chloride (PVC)	Versatili-ty, ease of blending, strength, toughness.	Blister packag-ing for non-food items; cling films for non-food use. May be used for food packaging with the addition of the plasticisers needed to make natively rigid PVC flexible . Non-packaging uses are electrical cable insulation; rigid piping; vinyl records.	Tm = 240; Tg = 85	2.4-4.1
04 PE-LD	Low-density polyethylene (LDPE)	Ease of processing, strength, toughness, flexibility, ease of seal-ing, barrier to moisture.	Frozen food bags; squeezable bottles, e.g. honey, mustard; cling films; flexible container lids.	Tm = 120; Tg = -125	0.17-0.28
05 PP	Polypropylene (PP)	Strength, toughness, resistance to heat, chemicals, grease and oil, versatile, barrier to moisture.	Reusable micro-waveable ware; kitchenware; yogurt containers; marga-rine tubs; micro-waveable disposable take-away con-tainers; disposable cups; soft drink bottle caps; plates.	Tm = 173; Tg = -10	1.5-2

♻ 06 PS	Polystyrene (PS)	Versatility, clarity, easily formed	Egg cartons; packing peanuts; disposable cups, plates, trays and cutlery; disposable take-away containers.	Tm = 240 (only isotactic); Tg = 100 (atactic and isotactic)	3-3.5
♻ 07 O	Other (often polycarbonate or ABS)	Dependent on polymers or combination of polymers	Beverage bottles; baby milk bottles. Non-packaging uses for polycarbonate: compact discs; "unbreakable" glazing; electronic apparatus housings; lenses including sunglasses, prescription glasses, automotive headlamps, riot shields, instrument panels.	Polycarbonate: Tg = 145; Tm = 225	Polycarbonate: 2.6; ABS plastics: 2.3

United States

Low national plastic recycling rates have been due to the complexity of sorting and processing, unfavorable economics, and consumer confusion about which plastics can actually be recycled. Part of the confusion has been due to the use of the resin identification code which is not on all plastic parts but just a subset that includes the recycling symbol as part of its design. The resin identification code is stamped or printed on the bottom of containers and surrounded by a triangle of arrows. The intent of these symbols was to make it easier to identify the type of plastics used to make a particular container and to indicate that the plastic is potentially recyclable. The question that remains is which types of plastics can be recycled by your local recycling center. In many communities, not all types of plastics are accepted for sidewalk recycling collection programs due to the high processing costs and complexity of the equipment required to recycle certain materials. There is also sometimes a seemingly low demand for the recycled product depending on a recycling center's proximity to entities seeking recycled materials. Another major barrier is that the cost to recycle certain materials and the corresponding market price for those materials sometimes does not present any opportunity for profit. The best example of this is polystyrene (commonly called styrofoam), although some communities, like Brookline, MA, are moving toward banning the distribution of polystyrene containers by local food and coffee businesses.

Recycling Codes

Recycling codes are used to identify the material from which an item is made, to facilitate easier recycling or other reprocessing. Having a recycling code, the chasing arrows logo or a resin code on an item is not an automatic indicator that a material is recyclable but rather an explanation of what the item is. Such symbols have been defined for batteries, biomatter/organic material, glass, metals, paper, and plastics. Various countries have adopted different codes. For example, the Table below shows the polymer resin codes (plastic) for a country. In the United States there are fewer as ABS is grouped in with other in group 7. Other countries have a more granular recycling code system. For example, China's polymer identification system has seven different classifications of plastic, five different symbols for post-consumer paths, and 140 identification codes The lack of codes in some countries has encouraged those who can fabricate their own plastic products, such as RepRap and other prosumer 3-D printer users, to adopt a voluntary recycling code based on the more comprehensive Chinese system.

List

Symbol	Code	Description	Examples
Plastics			
01 PET	#1 PET(E)	Polyethylene terephthalate	Polyester fibers, soft drink bottles
02 PE-HD	#2 PEHD or HDPE	High-density polyethylene	Plastic bottles, plastic bags, trash cans, imitation wood
03 PVC	#3 PVC	Polyvinyl chloride	Window frames, bottles for chemicals, flooring, plumbing pipes
04 PE-LD	#4 PELD or LDPE	Low-density polyethylene	Plastic bags, buckets, soap dispenser bottles, plastic tubes
05 PP	#5 PP	Polypropylene	Bumpers, car interior trim, industrial fibers, carry-out beverage cups
06 PS	#6 PS	Polystyrene	Toys, flower pots, video cassettes, ashtrays, trunks, beverage/food coolers, beer cups, wine and champagne cups, carry-out food containers, Styrofoam
07 O	#7 O (OTHER)	All other plastics	Polycarbonate (PC), polyamide (PA), styrene acrylonitrile (SAN), acrylic plastics/polyacrylonitrile (PAN), bioplastics
ABS	#9 or #ABS	Acrylonitrile butadiene styrene	Monitor/TV cases, coffee makers, cell phones, most computer plastic

	PA	Polyamide	Nylon
		Batteries	
	#8 Lead	Lead–acid battery	Car batteries
	#9 or #19 Alkaline	Alkaline battery	
	#10 NiCD	Nickel–cadmium battery	
	#11 NiMH	Nickel–metal hydride battery	
	#12 Li	Lithium battery	
	#13 SO(Z)	Silver-oxide battery	
	#14 CZ	Zinc–carbon battery	
		Paper	
	#20 C PAP (PCB)	Cardboard	
	#21 PAP	Other paper	Mixed paper magazines, mail
	#22 PAP	Wax Paper (single sided)	McDonald's: fast food sandwich wrappers, meat packing, gum wrappers, some drink boxes, BetaMax boxes.
	#23 PBD (PPB)	Paperboard	Greeting cards, frozen food boxes, book covers
		Metals	
	#40 FE	Steel	
	#41 ALU	Aluminium	
		Biomatter/Organic material	
	#50 FOR	Wood	
	#51 FOR	Cork	Bottle stoppers, place mats, construction material

Symbol	Code	Material	Uses
△ 60 TEX	#60 COT	Cotton	
△ 61 TEX	#61 TEX	Jute	
	#62-69 TEX	Other Textiles	
Glass			
△ 70 GL	#70 GLS	Mixed Glass Container/ Multi-Part Container	
△ 71 GL	#71 GLS	Clear Glass	
△ 72 GL	#72 GLS	Green Glass	
	#73 GLS	Dark Sort Glass	
	#74 GLS	Light Sort Glass	
	#75 GLS	Light Leaded Glass	Televisions, high-end electronics display glass
	#76 GLS	Leaded Glass	Older televisions, ash trays, older beverage holders
	#77 GLS	Copper Mixed / Copper Backed Glass	Electronics, LCD display heads, clocks, watches
	#78 GLS	Silver Mixed / Silver Backed Glass	Mirrors, formal table settings
	#79 GLS	Gold Mixed / Gold Backed Glass	Computer glass, formal table settings
Composites (80—99)			
	#81 PapPet	Paper + plastic	Consumer packaging, pet food bags, cold store grocery bags, Icecream containers, cardboard cans, disposable plates
	#82	Paper and fibreboard / Aluminium	
	#83	Paper and fibreboard / Tinplate	
△ 84 C/PAP	#84 C/PAP (or PapAl)	Paper and cardboard / Plastic / Aluminium	Liquid storage containers, juice boxes, cardboard cans, cigarette pack liners, gum wrappers, cartage shells for blanks, fire-works colouring material, Tetra Brik.

	#85	Paper and fibreboard / plastic / aluminium / tinplate	
#87 Card-stock Laminate	Biodegradable plastic	Laminating material, special occasion cards, bookmarks, business cards, flyers/advertising	
	#90	Plastics / Aluminium	
	#91	Plastic / Tinplate	
	#92	Plastic / Miscellaneous metals	
	#95	Glass / Plastic	
	#96	Glass / Aluminium	
	#97	Glass / Tinplate	
	#98	Glass / Miscellaneous metals	

Battery Recycling

Battery recycling is a recycling activity that aims to reduce the number of batteries being disposed as municipal solid waste. Batteries contain a number of heavy metals and toxic chemicals and disposing them by the same process as regular trash has raised concerns over soil contamination and water pollution.

Li-ion

Battery Recycling by Type

Most types of batteries can be recycled. However, some batteries are recycled more readily than others, such as lead–acid automotive batteries (nearly 90% are recycled) and button cells (because of the value and toxicity of their chemicals). Other types, such as alkaline and rechargeable, e.g., nickel–cadmium (Ni-Cd), nickel metal hydride (Ni-MH), lithium-ion (Li-ion) and nickel–zinc (Ni-Zn), can also be recycled.

Lead–acid Batteries

These batteries include but are not limited to: car batteries, golf cart batteries, UPS batteries, industrial fork-lift batteries, motorcycle batteries, and commercial batteries.

These can be regular lead–acid, sealed lead–acid, gel type, or absorbed glass mat batteries. These are recycled by grinding them, neutralizing the acid, and separating the polymers from the lead. The recovered materials are used in a variety of applications, including new batteries.

Recycling the lead from batteries.

The lead in a lead–acid battery can be recycled. Elemental lead is toxic and should therefore be kept out of the waste stream.

Lead–acid batteries collected by an auto parts retailer for recycling.

Many cities offer battery recycling services for lead–acid batteries. In some jurisdictions, including U.S. states and Canadian provinces, a refundable deposit is paid on batteries. This encourages recycling of old batteries instead of abandonment or disposal with household waste. In the United States, about 99% of lead from used batteries is reclaimed.

Businesses that sell new car batteries may also collect used batteries (or be required to do so by law) for recycling. Some businesses accept old batteries on a "walk-in" basis, as opposed to in exchange for a new battery. Most battery shops and recycling centres pay for scrap batteries. This can be a lucrative business, enticing especially to risk-takers because of the wild fluctuations in the value of scrap lead that can occur overnight. When lead prices go up, scrap batteries become targets for thieves.

Silver Oxide Batteries

Used most frequently in watches, toys, and some medical devices, silver oxide batteries contain a small amount of mercury. Most jurisdictions regulate their handling and disposal to reduce the discharge of mercury into the environment. Silver oxide batteries can be recycled to recover the mercury.

Lithium Ion Batteries

Lithium-ion batteries and lithium iron phosphate (LiFePO4) batteries often contain among other useful metals high-grade copper and aluminium in addition to – depending on the active material – transition metals cobalt and nickel as well as rare earths. To prevent a future shortage of cobalt, nickel, and lithium and to enable a sustainable life cycle of these technologies, recycling processes for lithium batteries are needed. These processes have to regain not only cobalt, nickel, copper, and aluminum from spent battery cells, but also a significant share of lithium. In order to achieve this goal, several unit operations are combined to complex process chains, especially considering the task to recover high rates of valuable materials with regard to involved safety issues.

These unit operations are:

- Deactivation or discharging of the battery (especially in case of batteries from electric vehicles)

- Disassembly of battery systems (especially in case of batteries from electric vehicles)

- Mechanical Processes (including crushing, sorting, and sieving processes)

- Hydrometallurgical processes

- Pyrometallurgical processes

Specific dangers associated with lithium-ion battery recycling processes are: electrical dangers, chemical dangers, burning reactions, and their potential interactions. A complicating factor is the water sensitivity: lithium hexafluorophosphate, a possible electrolyte material, will react with water to form hydrofluoric acid; cells are often immersed in a solvent to prevent this. Once removed, the jelly rolls are separated and the materials removed by ultrasonic agitation, leaving the electrodes ready for melting down and recycling.

Pouch cells are particularly easier to recycle in this way and some people already do this to salvage the copper despite the safety issues.

Battery Composition by Type

Italics designates button cell types.
Bold designates secondary types.
All figures are percentages; due to rounding they may not add up to exactly 100.

Type	Fe	Mn	Ni	Zn	Hg	Li	Ag	Cd	Co	Al	Pb	Other	KOH	Paper	Plastic	Alkali	C	Acids	Water	Other
Alkaline	24.8	22.3	0.5	14.9								1.3		1	2.2	5.4	3.7		10.1	14
Zinc-carbon	16.8	15		19.4							0.1	0.8		0.7	4	6	9.2		12.3	15.2
Lithium	50	19	1			2									7		2			19
Mercury-oxide	37	1	1	14	31								2		3		1		3	7
Zinc-air	42			35	1										4	4	1		10	3
Lithium	60	18	1			3									3		2			13
Alkaline	37	23	1	11	0.6										6	2	2		6	14
Silver oxide	42	2	2	9	0.4		31					4			2	1	0.5		2	4
Nickel-cadmium	35		22					15							10	2			5	11
NiMH	20	1	35	1					4			10			9	4			8	8
Li-ion	22					3			18	5		11					13			28
Lead–acid											65	4			10			16		5

Battery Recycling by Location

Several sizes of button and coin cell. They are all recyclable in the UK and Ireland.

Country	Return percentage	
	2002	2012
🇨🇭 Switzerland	61 %	73 %
🇧🇪 Belgium	59 %	-
🇸🇪 Sweden	55 %	-
🇩🇪 Germany	39 %	44 %
🇦🇹 Austria	44 %	-
🇳🇱 Netherlands	32 %	-
🇬🇧 United Kingdom	-	32 %
🇫🇷 France	16 %	-
🇨🇦 Canada	3 %	5.6 %

* Figures for Q1 and Q2 2012.

European Union

In 2006 the EU passed the Battery Directive of which one of the aims is a higher rate of battery recycling. The EU directive states that at least 25% of all the EU's used batteries must be collected by 2012, and rising to no less than 45% by 2016, of which, that at least 50% of them must be recycled.

Channel Islands

In early 2009 Guernsey took the initiative by setting up the Longue Hougue recycling facility, which, among other functions, offers a drop-off point for used batteries so they can be recycled off island. The resulting publicity meant that a lot of people complied with the request to dispose of batteries responsibly.

United Kingdom

From April 2005 to March 2008, the UK non-governmental body WRAP conducted trials of battery recycling methods around the UK. The methods tested were: Kerbside, retail drop-off, community drop-off, postal, and hospital and fire station trials. The kerbside trials collected the most battery mass, and were the most well-received and understood by the public. The community drop-off containers which were spread around local community areas were also relatively successful in terms of mass of batteries collected. The lowest performing were the hospital and fire service trials (although these served their purpose very well for specialized battery types like hearing aid and smoke alarm batteries). Retail drop off trials were the second most effective (by volume) method but one of the least well received and used by the public. Both the kerbside and postal trials received the highest awareness and community support.

Household batteries can be recycled in United Kingdom at council recycling sites as well as at some shops and shopping centres—e.g., Dixons, Currys, The Link and PC World.

A scheme started in 2008 by a large retail company allowed household batteries to be posted free of charge in envelopes available at their shops. This scheme was cancelled at the request of the Royal Mail because of hazardous industrial battery waste being sent as well as household batteries.

An EU directive on batteries that came into force in 2009 means producers must pay for the collection, treatment, and recycling of batteries. This has yet to be ratified into UK law however, so there is currently no real incentive for producers to provide the necessary services.

From 1 February 2010 batteries can be recycled anywhere the Be Positive sign appears. Shops and online retailers that sell more than 32 kilograms of batteries a year must offer facilities to recycle batteries. This is equivalent to one pack of four AA batteries a day. Shops which sell this amount must by law provide recycling facilities as of 1 February 2010.

In Great Britain an increasing number of shops (Argos, Homebase, B&Q, Tesco, and Sainsbury's) are providing battery return boxes and cylinders for their customers.

North America

The rechargeable battery industry has formed the Rechargeable Battery Recycling Corporation (RBRC), which operates a free battery recycling program called Call2Recycle throughout the United States and Canada. RBRC provides businesses with prepaid shipping containers for rechargeable batteries of all types while consumers can drop off batteries at numerous participating collection centers. It claims that no component of any recycled battery eventually reaches a landfill.

A study estimated battery recycling rates in Canada based on RBRC data. In 2002, it wrote, the collection rate was 3.2%. This implies that 3.2% of rechargeable batteries were recycled, and the rest were thrown in the trash. By 2005, it concluded, the collection rate had risen to 5.6%.

In 2009, Kelleher Environmental updated the study. The update estimates the following. "Collection rate values for the 5 [and] 15 year hoarding assumptions respectively are: 8% to 9% for NiCd batteries; 7% to 8% for NiMH batteries; and 45% to 72% for lithium ion and lithium polymer batteries combined. Collection rates through the [RBRC] program for all end of life small sealed lead acid (SLA) consumer batteries were estimated at 10% for 5 year and 15 year hoarding assumptions. [...] It should also be stressed that these figures do not take collection of secondary consumer batteries through other sources into account, and actual collection rates are likely higher than these values."

A November 2011 *New York Times* article reported that batteries collected in the United States are increasingly being transported to Mexico for recycling as a result of a widening gap between the strictness of environmental and labor regulations between the two countries.

In 2015, Energizer announced availability of disposable AAA and AA alkaline batteries made with 3.8% to 4% (by weight) of recycled batteries, branded as EcoAdvanced.

Japan

Japan does not have a single national battery recycling law, so the advice given is to follow local and regional statutes and codes in disposing batteries. The Battery Association of Japan (BAJ) recommends that alkaline, zinc-carbon, and lithium primary batteries can be disposed of as normal household waste. The BAJ's stance on button cell and secondary batteries is toward recycling and increasing national standardisation of procedures for dealing with these types of batteries.

In April 2004 the Japan Portable Rechargeable Battery Recycling Center (JBRC) was created to handle and promote battery recycling throughout Japan. They provide battery recycling containers to shops and other collection points.

Paint Recycling

Paint is a recyclable item. Latex paint is collected at collection facilities in many countries and shipped to paint-recycling facilities.

How Paint is Recycled

There are many ways that paint can be recycled. Most often, the highest quality of latex paint is sorted out and turned back into recycled paint that can be used. Recycled paint is environmentally preferable to new paint, while still maintaining comparable quality. In many cases, reusable paints of the same color are pumped into a tank where the material is mixed and tested. The paint is adjusted with additives and colorants as necessary. Finally, the paint is fine filtered and packaged for sale.

Paint that cannot be reused has other environmentally friendly uses. Non-reusable paint can be made into a product used in cement manufacturing, thereby recycling virtually 100% of the original paint.

Recycling one gallon of paint could save 13 gallons of water, and 250,000 gallons of water pollution

Paint Product Stewardship Initiative

Paint Recycling by Country

Canada

In Ontario, Stewardship Ontario oversees the collection of waste paint from consumers and diversion from landfill to meet targets approved by the Ministry of the Environment through a program called the Orange Drop Program. The Orange Drop program is an extensive and growing network of collection sites—drop-off locations for paint leftovers and other special materials that can't go in the Blue Box or the garbage.

As an Orange Drop-approved transporter and processor, Loop Recycled Products Inc. takes leftover paint, collected through Stewardship Ontario, and turns it into 12 shades of premium, affordable and environmentally friendly recycled paint. Reusing top-quality residual paint (on average, the original retail value of a gallon of incoming paint is approximately $30) enables Loop to create premium products without the raw material costs and energy consumption needed to make paint from scratch.

Since 2012, Loop Recycled Products Inc. has diverted over 6 million litres of paint from disposal in Ontario's landfills, incineration and waterways and is committed to innovation and solving Canada's waste paint problem.

In February 2015 Waste Diversion Ontario approved Product Care as the new Ontario waste paint stewardship operator effectively replacing Stewardship Ontario.

Alberta's paint recycling program started accepting leftover, unwanted paint on April 1, 2008. It is estimated that about 30 million liters of paint is sold in Alberta each year. On average, 5 to 10 percent of this ends up as waste, which can pose environmental and health risks if disposed of improperly. Paint contains many components that have great potential for reuse, recycling and recovery. The Paint Recycling Alberta program enables these products to be handled and recycled in an environmentally safe manner, reducing their impact on the environment. The program is funded through environmental fees charged on the sale of new paint in Alberta. The fees are put into a dedicated fund that can only be used to manage the paint recycling program.

The paint is sorted into different streams and sent to registered processors to be recycled into new paint, used in other products or in energy recovery, or sent for proper disposal if necessary. Any processor that receives paint must be registered with the Paint Recycling Program and meet all applicable environmental, transportation, health & safety, and local requirements.

Calibre Environmental LTD. (CEL) located in Calgary, Alberta, became a key part in 2008 of the new Alberta Paint Stewardship program which significantly increased the recycling of unused latex paint from across the province of Alberta. Calibre Environmental Ltd. currently processes about 1.6 million kilograms of latex paint annually, which equates to the successful recycling of one million litres of quality latex paint per year.

New Zealand

United Kingdom

In the UK reusable leftover paint can be donated to Community RePaint, a national network of paint reuse schemes. The network comprises local schemes run by not-for-profit organisations, local authorities or waste management companies, in the Community RePaint network. The schemes collect surplus paint from trade sources i.e. painters, decorators, retailers, manufacturers, and/or leftover paint donated by householders at council household waste and recycling centres (also known as tips). The paint is then sorted by staff and volunteers before being redistributed to local charities, community groups, families and individuals in need. The Community RePaint network, is sponsored by Dulux (part of AkzoNobel), managed by an environmental consultancy, Resource Futures and has been cited as an example of best practice for the management of surplus paint in a report by the European Commission and by DEFRA in Guidance on Applying the Waste Hierarchy.

There are also a handful of companies recycling paint in the UK, but only two reprocessing waste paint, back into the quality of virgin paint; Newlife Paints. Newlife Paints was formed in 2008 after Keith Harrison, an industrial chemist, developed a process that converted waste emulsion paint back into full quality, commercial grade paint. Castle RePaint, part of the social enterprise company Castle Furniture also consolidates unwanted emulsion paint into brand new 'RePaint' in a range of colours.

United States

Concerns about the life cycle of paint have led to the creation of PaintCare, a non-profit 501(c)(3) organization established to represent paint manufacturers (paint producers) to plan and operate paint stewardship programs in the United States in those states that pass paint stewardship laws.

Paint stewardship law aims to enable the paint industry to implement a collection program that allows consumers to take their leftover, unwanted paint to a collection site to be collected and recycled. Legislation mandating the creation of the PaintCare program has been enacted in eight states since 2009: Oregon, California, Connecticut, Rhode Island, Vermont, Minnesota, Maine, and Colorado. Legislation has also been passed for the District of Columbia; PaintCare anticipates beginning the District's paint stewardship program in September 2016. PaintCare is responsible for promoting the reuse of post-consumer architectural paint (leftover paint) and providing for the collection, transport, and processing of this paint using the hierarchy of "reduce, reuse, recycle," and proper disposal. Most PaintCare locations are at paint retailers who volunteer to take back paint. These retailers take back paint during regular business hours, making paint recycling and disposal much more convenient for the public.

Paint is shipped to companies such as Amazon Environmental, GDB International, Metro Paint (Oregon), UCI Environmental (Nevada) and Kelly Moore, Visions Paint Recycling, Inc (California)& Williams Paint Recycling Company. In the Southern California area, Acrylatex Coatings & Recycling, Inc. accepts unused/unwanted latex paints for reprocessing into a viable resource of recycled paints in 20-standard colors. In the southeastern United States Atlanta Paint Disposal has a paint recycling program with drop off locations in Atlanta, Georgia.

A new charitable organization known as The Global Paint for Charity incorporated in Georgia, US, has as its mission to collect leftover paint from residents and businesses nationwide and use it for global housing rehabilitation projects, including homes, schools, hospitals, jails and churches for vulnerable families in developing countries. They partner with non-profit organizations with existing operations in these continents for paint distribution. Through the support of their donors and partners they are able to improve communities, increase access to quality paints and protect the environment.

The Environmental Protection Agency (EPA) estimates that every homeowner in the US has 3 to 4 gallons of leftover paints in their basement, and 10 percent of those paints ends up in landfills. One gallon of improperly disposed paint has the ability to pollute up to 250,000 gallons of water. By participating in the program, individuals and businesses will take greater steps to protect the environment, and improve living conditions for vulnerable populations throughout the world. If you would like to support the Global Paint for Charity, they encourage you to take action today.

Improve access of high-quality paints to vulnerable populations around the world. Nearly 2.5 billion people in developing countries live on less than $2 a day. For them, paint is very expensive. In these settings, it is very difficult for families to secure sufficient income for their basic needs (including but not limited to: food, medicine, water, clothes, school supplies, and shelter). When making consumption choices that involve spending on their basic needs there is nothing left to spend on paint. The paint shortage affects many other areas of the world, where communities lack even the most basic need and materials to uplift their people. For the world's poorest communities, home isn't just where the heart is. Dirt walls and neglected communities are not attractive to tourists, putting those who can't afford paint, not only at the greatest rick of life-threatening of bad germs but also lack of economic opportunities.

Since it started years ago, as many as 500–6000 gallons of paint have been shipped at a time to developing countries, including Kenya, Uganda, Haiti, Dominican Republic, Honduras, El Salvador, Guyana, Guinea, Ghana, Jamaica and Mexico, 240 volunteers have painted 459 family homes and 40 schools and orphanages with over 150,000 gallons of donated paint from businesses and residences.

Global Paint for Charity continues to expand and impact our communities around the world. They do this through working directly with their volunteers, donors, and partners. From developing paint projects that engage their employees in the beautification of the community; to run frequent local paint drives to support their program.

Concrete Recycling

Concrete from a building being sent to a portable crusher. This is the first step to recycling concrete.

When structures made of concrete are demolished or renovated, concrete recycling is an increasingly common method of utilizing the rubble. Concrete was once routinely trucked to landfills for disposal, but recycling has a number of benefits that have made it a more attractive option in this age of greater environmental awareness, more environmental laws, and the desire to keep construction costs down.

Concrete aggregate collected from demolition sites is put through a crushing machine. Crushing facilities accept only uncontaminated concrete, which must be free of trash, wood, paper and other such materials. Metals such as rebar are accepted, since they can be removed with magnets and other sorting devices and melted down for recycling elsewhere. The remaining aggregate chunks are sorted by size. Larger chunks may go through the crusher again. After crushing has taken place, other particulates are filtered out through a variety of methods including hand-picking and water flotation.

Crushing at the actual construction site using portable crushers reduces construction costs and the pollution generated when compared with transporting material to and from a quarry. Large road-portable plants can crush concrete and asphalt rubble at 600 tons per hour or more. These systems normally consist of a rubble crusher, side discharge conveyor, screening plant, and a return conveyor from the screen to the crusher inlet for reprocessing oversize materials. Compact, self-contained mini-crushers are also available that can handle up to 150 tons per hour and fit into tighter areas. With the advent of crusher attachments - those connected to various construction equipment, such as excavators - the trend towards recycling on-site with smaller volumes of material is growing rapidly. These attachments encompass volumes of 100 tons/hour and less.

Uses of Recycled Concrete

Smaller pieces of concrete are used as gravel for new construction projects. Sub-base gravel is laid down as the lowest layer in a road, with fresh concrete or asphalt poured over it. The US Federal Highway Administration may use techniques such as these to build new highways from the materials of old highways. Crushed recycled concrete can also be used as the dry aggregate for brand new concrete if it is free of contaminants. Also, concrete pavements can be broken in place and used as a base layer for an asphalt pavement through a process called rubblization.

Larger pieces of crushed concrete can be used as riprap revetments, which are "a very effective and popular method of controlling streambank erosion."

With proper quality control at the crushing facility, well graded and aesthetically pleasing materials can be provided as a substitute for landscaping stone or mulch.

Wire gabions (cages), can be filled with crushed concrete and stacked together to provide economical retaining walls. Stacked gabions are also used to build privacy screen walls (in lieu of fencing).

Benefits

There are a variety of benefits in recycling concrete rather than dumping it or burying it in a landfill.

- Keeping concrete debris out of landfills saves landfill space.

- Using recycled material as gravel reduces the need for gravel mining.

- Using recycled concrete as the base material for roadways reduces the pollution involved in trucking material.

- Recycling one ton of cement could save 1,360 gallons water, 900 kg of CO_2

Lead Paint Contamination

There have been concerns about the recycling of painted concrete due to possible lead content. The Army Corps of Engineers' Construction Engineering Research Laboratory (CERL) and others have conducted studies to see if lead-based paint in crushed concrete actually poses a hazard. It was concluded that concrete with lead-based paint would be able to be used as clean fill without impervious cover but with some type of soil cover.

References

- Cleveland, Cutler J.; Morris, Christopher G. (November 15, 2013). Handbook of Energy: Chronologies, Top Ten Lists, and Word Clouds. Elsevier. p. 461. ISBN 978-0-12-417019-3.

- Dadd-Redalia, Debra (January 1, 1994). Sustaining the earth: choosing consumer products that are safe for you, your family, and the earth. New York: Hearst Books. p. 103. ISBN 978-0-688-12335-2.

- Carl A. Zimring (2005). Cash for Your Trash: Scrap Recycling in America. New Brunswick, NJ: Rutgers University Press. ISBN 0-8135-4694-X.

- Lynn R. Kahle; Eda Gurel-Atay, eds. (2014). Communicating Sustainability for the Green Economy. New York: M.E. Sharpe. ISBN 978-0-7656-3680-5.

- Huesemann, M.; Huesemann, J. (2011). Techno-fix: Why Technology Won't Save Us or the Environment. New Society Publishers. p. 464. ISBN 978-0-86571-704-6. Retrieved 2016-07-07.

- Foster, J. B.; Clark, B. (2011). The Ecological Rift: Capitalisms War on the Earth. Monthly Review Press. p. 544. ISBN 1-58367-218-4.

- Ministry for the Environment (December 2007). Environment New Zealand 2007. Ministry for the Environment (New Zealand). ISBN 978-0-478-30192-2. Retrieved 2008-03-27.

- Nguemaleu, Raoul-Abelin Choumin; Montheu, Lionel (2014-05-09). Roadmap to Greener Computing. CRC Press. p. 170. ISBN 9781466506848.

- The Self-Sufficiency Handbook: A Complete Guide to Greener Living by Alan Bridgewater pg. 62--Skyhorse Publishing Inc., 2007 ISBN 1-60239-163-7, ISBN 978-1-60239-163-5

- "Report: "On the Making of Silk Purses from Sows' Ears," 1921: Exhibits: Institute Archives & Special Collections: MIT". mit.edu. Retrieved July 7, 2016.

- Steven E. Landsburg. "Why I Am Not An Environmentalist: The Science of Economics Versus the Religion of Ecology Excerpt from The Armchair Economist: Economics & Everyday Life" (PDF) (PDF). Retrieved July 6, 2016.

- UNFCCC (2007). "Investment and financial flows to address climate change" (PDF). unfccc.int. UNFCCC. p. 81. Retrieved 2016-07-07.

Oil Spill and Separation

Oil spills are a common and extremely destructive form of man-made industrial accident. An API oil-water separator is a device designed to separate huge amounts of oil from the wastewater. There are other devices also used for separating oil such as the hydrocyclone. The chapter discusses the methods of oil separation in a critical manner providing key analysis to the subject matter.

Oil Spill

An oil spill is the release of a liquid petroleum hydrocarbon into the environment, especially marine areas, due to human activity, and is a form of pollution. The term is usually applied to marine oil spills, where oil is released into the ocean or coastal waters, but spills may also occur on land. Oil spills may be due to releases of crude oil from tankers, offshore platforms, drilling rigs and wells, as well as spills of refined petroleum products (such as gasoline, diesel) and their by-products, heavier fuels used by large ships such as bunker fuel, or the spill of any oily refuse or waste oil.

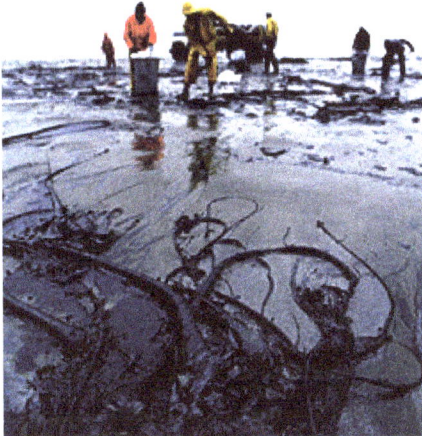

Help after an oil spill

Oil spills penetrate into the structure of the plumage of birds and the fur of mammals, reducing its insulating ability, and making them more vulnerable to temperature fluctuations and much less buoyant in the water. Cleanup and recovery from an oil spill is difficult and depends upon many factors, including the type of oil spilled, the temperature of the water (affecting evaporation and biodegradation), and the types of shorelines and beaches involved. Spills may take weeks, months or even years to clean up.

Oil slick from the Montara oil spill in the Timor Sea, September 2009

Oil spills can have disastrous consequences for society; economically, environmentally, and socially. As a result, oil spill accidents have initiated intense media attention and political uproar, bringing many together in a political struggle concerning government response to oil spills and what actions can best prevent them from happening.

Largest Oil Spills

Crude oil and refined fuel spills from tanker ship accidents have damaged vulnerable ecosystems in Alaska, the Gulf of Mexico, the Galapagos Islands, France, the Sundarbans, Ogoniland, and many other places. The quantity of oil spilled during accidents has ranged from a few hundred tons to several hundred thousand tons (e.g., Deepwater Horizon Oil Spill, Atlantic Empress, Amoco Cadiz), but volume is a limited barometer of damage or impact. Smaller spills have already proven to have a great impact on ecosystems, such as the Exxon Valdez oil spill because of the remoteness of the site or the difficulty of an emergency environmental response.

Oil spills at sea are generally much more damaging than those on land, since they can spread for hundreds of nautical miles in a thin oil slick which can cover beaches with a thin coating of oil. These can kill seabirds, mammals, shellfish and other organisms they coat. Oil spills on land are more readily containable if a makeshift earth dam can be rapidly bulldozed around the spill site before most of the oil escapes, and land animals can avoid the oil more easily.

1. One metric ton (tonne) of crude oil is roughly equal to 308 US gallons or 7.33 barrels approx.; 1 oil barrel (bbl) is equal to 35 imperial or 42 US gallons. Approximate conversion factors.

2. Estimates for the amount of oil burned in the Kuwaiti oil fires range from 500,000,000 barrels (79,000,000 m³) to nearly 2,000,000,000 barrels (320,000,000 m³). Between 605 and 732 wells were set ablaze, while many others were severely damaged and gushed uncontrolled for several months. It took over ten months to bring all of the wells under control. The fires alone were estimated to consume approximately 6,000,000 barrels (950,000 m³) of oil per day at their peak.

Largest oil spills

Spill / Tanker	Location	Date	Tonnes of crude oil (thousands)[a]	Barrels (thousands)	US Gallons (thousands)	References
Kuwaiti oil fires[dubious – discuss][b]	Kuwait	January 16, 1991 - November 6, 1991	136,000	1,000,000	42,000,000	
Kuwaiti oil lakes[c]	Kuwait	January 1991 - November 1991	3,409–6,818	25,000–50,000	1,050,000–2,100,000	
Lakeview Gusher	United States, Kern County, California	March 14, 1910 – September 1911	1,200	9,000	378,000	
Gulf War oil spill[d]	Kuwait, Iraq, and the Persian Gulf	January 19, 1991 - January 28, 1991	818–1,091	6,000–8,000	252,000–336,000	
Deepwater Horizon	United States, Gulf of Mexico	April 20, 2010 – July 15, 2010	560–585	4,100–4,900	172,000–180,800	
Ixtoc I	Mexico, Gulf of Mexico	June 3, 1979 – March 23, 1980	454–480	3,329–3,520	139,818–147,840	
Atlantic Empress / Aegean Captain	Trinidad and Tobago	July 19, 1979	287	2,105	88,396	
Fergana Valley	Uzbekistan	March 2, 1992	285	2,090	87,780	
Nowruz Field Platform	Iran, Persian Gulf	February 4, 1983	260	1,907	80,080	
ABT Summer	Angola, 700 nmi (1,300 km; 810 mi) offshore	May 28, 1991	260	1,907	80,080	
Castillo de Bellver	South Africa, Saldanha Bay	August 6, 1983	252	1,848	77,616	
Amoco Cadiz	France, Brittany	March 16, 1978	223	1,635	68,684	

3. Oil spilled from sabotaged fields in Kuwait during the 1991 Persian Gulf War pooled in approximately 300 oil lakes, estimated by the Kuwaiti Oil Minister to contain approximately 25,000,000 to 50,000,000 barrels (7,900,000 m³) of oil. According to the U.S. Geological Survey, this figure does not include the amount of oil absorbed by the ground, forming a layer of "tarcrete" over approximately five percent of the surface of Kuwait, fifty times the area occupied by the oil lakes.

4. Estimates for the Gulf War oil spill range from 4,000,000 to 11,000,000 barrels (1,700,000 m³). The figure of 6,000,000 to 8,000,000 barrels (1,300,000 m³) is the range adopted by the U.S. Environmental Protection Agency and the United Nations in the immediate aftermath of the war, 1991–1993, and is still current, as cited by NOAA and The New York Times in 2010. This amount only includes oil discharged directly into the Persian Gulf by the retreating Iraqi forces from January 19 to 28, 1991. However, according to the U.N. report, oil from other sources not included in the official estimates continued to pour into the Persian Gulf through June, 1991. The amount of this oil was estimated to be at least several hundred thousand barrels, and may have factored into the estimates above 8,000,000 barrels (1,300,000 m³).

Human Impact

An oil spill represents an immediate fire hazard. The Kuwaiti oil fires produced air pollution that caused respiratory distress. The Deepwater Horizon explosion killed eleven oil rig workers. The fire resulting from the Lac-Mégantic derailment killed 47 and destroyed half of the town's centre.

Spilled oil can also contaminate drinking water supplies. For example, in 2013 two different oil spills contaminated water supplies for 300,000 in Miri, Malaysia; 80,000 people in Coca, Ecuador,. In 2000, springs were contaminated by an oil spill in Clark County, Kentucky.

Contamination can have an economic impact on tourism and marine resource extraction industries. For example, the Deepwater Horizon oil spill impacted beach tourism and fishing along the Gulf Coast, and the responsible parties were required to compensate economic victims.

Environmental Effects

In general, spilled oil can affect animals and plants in two ways: direct from the oil and from the response or cleanup process. There is no clear relationship between the amount of oil in the aquatic environment and the likely impact on biodiversity. A smaller spill at the wrong time/wrong season and in a sensitive environment may prove much more harmful than a larger spill at another time of the year in another or even the same environment. Oil penetrates into the structure of the plumage of birds and the fur of mammals, reducing its insulating ability, and making them more vulnerable to temperature fluctuations and much less buoyant in the water.

A surf scoter covered in oil as a result of the 2007 San Francisco Bay oil spill

Animals who rely on scent to find their babies or mothers cannot due to the strong scent of the oil. This causes a baby to be rejected and abandoned, leaving the babies to starve and eventually die. Oil can impair a bird's ability to fly, preventing it from foraging or escaping from predators. As they preen, birds may ingest the oil coating their feathers, irritating the digestive tract, altering liver function, and causing kidney damage. Together with their diminished foraging capacity, this can rapidly result in dehydration and metabolic imbalance. Some birds exposed to petroleum also experience changes in their hormonal balance, including changes in their luteinizing protein. The majority of birds affected by oil spills die from complications without human intervention. Some studies have suggested that less than one percent of oil-soaked birds survive, even after cleaning, although the survival rate can also exceed ninety percent, as in the case of the Treasure oil spill.

A bird covered in oil from the Black Sea oil spill

Heavily furred marine mammals exposed to oil spills are affected in similar ways. Oil coats the fur of sea otters and seals, reducing its insulating effect, and leading to fluctuations in body temperature and hypothermia. Oil can also blind an animal, leaving it defenseless. The ingestion of oil causes dehydration and impairs the digestive process. Animals can be poisoned, and may die from oil entering the lungs or liver.

There are three kinds of oil-consuming bacteria. Sulfate-reducing bacteria (SRB) and acid-producing bacteria are anaerobic, while general aerobic bacteria (GAB) are aerobic. These bacteria occur naturally and will act to remove oil from an ecosystem, and their biomass will tend to replace other populations in the food chain.

Sources and Rate of Occurrence

A VLCC tanker can carry 2 million barrels (320,000 m³) of crude oil. This is about eight times the amount spilled in the widely known *Exxon Valdez* incident. In this spill, the ship ran aground and dumped 260,000 barrels (41,000 m³) of oil into the ocean in March 1989. Despite efforts of scientists, managers, and volunteers over 400,000 seabirds, about 1,000 sea otters, and immense numbers of fish were killed. Considering the volume of oil carried by sea, however, tanker owners' organisations often argue that the industry's safety record is excellent, with only a tiny fraction of a percentage of oil cargoes carried ever being spilled. The International Association of Independent Tanker Owners has observed that "accidental oil spills this decade have been at record low levels—one third of the previous decade and one tenth of the 1970s—at a time when oil transported has more than doubled since the mid 1980s."

Oil tankers are only one source of oil spills. According to the United States Coast Guard, 35.7% of the volume of oil spilled in the United States from 1991 to 2004 came from tank vessels (ships/barges), 27.6% from facilities and other non-vessels, 19.9% from non-tank vessels, and 9.3% from pipelines; 7.4% from mystery spills. On the other hand, only 5% of the actual spills came from oil tankers, while 51.8% came from other kinds of vessels.

The International Tanker Owners Pollution Federation has tracked 9,351 accidental spills that have occurred since 1974. According to this study, most spills result from routine operations such as loading cargo, discharging cargo, and taking on fuel oil. 91% of the operational oil spills are small, resulting in less than 7 metric tons per spill. On the other hand, spills resulting from accidents like collisions, groundings, hull failures, and explosions are much larger, with 84% of these involving losses of over 700 metric tons.

Cleanup and Recovery

Cleanup and recovery from an oil spill is difficult and depends upon many factors, including the type of oil spilled, the temperature of the water (affecting evaporation and biodegradation), and the types of shorelines and beaches involved.

A U.S. Air Force Reserve plane sprays Corexit dispersant over the Deepwater Horizon oil spill in the Gulf of Mexico.

Clean-up efforts after the Exxon Valdez oil spill.

A US Navy oil spill response team drills with a "Harbour Buster high-speed oil containment system".

Methods for cleaning up include:

- Bioremediation: use of microorganisms or biological agents to break down or remove oil; such as the bacteria Alcanivorax or Methylocella Silvestris.

- Bioremediation Accelerator: Oleophilic, hydrophobic chemical, containing no bacteria, which chemically and physically bonds to both soluble and insoluble hydrocarbons. The bioremediation accelerator acts as a herding agent in water and on the surface, floating molecules to the surface of the water, including solubles such as phenols and BTEX, forming gel-like agglomerations. Undetectable levels of hydrocarbons can be obtained in produced water and manageable water columns. By overspraying sheen with bioremediation accelerator, sheen is eliminated within minutes. Whether applied on land or on water, the nutrient-rich emulsion creates a bloom of local, indigenous, pre-existing, hydrocarbon-consuming bacteria. Those specific bacteria break down the hydrocarbons into water and carbon dioxide, with EPA tests showing 98% of alkanes biodegraded in 28 days; and aromatics being biodegraded 200 times faster than in nature they also sometimes use the hydrofireboom to clean the oil up by taking it away from most of the oil and burning it.

- Controlled burning can effectively reduce the amount of oil in water, if done properly. But it can only be done in low wind, and can cause air pollution.

Oil slicks on Lake Maracaibo

Volunteers cleaning up the aftermath of the Prestige oil spill

- Dispersants can be used to dissipate oil slicks. A dispersant is either a non-surface active polymer or a surface-active substance added to a suspension, usually a colloid, to improve the separation of particles and to prevent settling or clumping. They may rapidly disperse large amounts of certain oil types from the sea surface by transferring it into the water column. They will cause the oil slick to break up and form water-soluble micelles that are rapidly diluted.

The oil is then effectively spread throughout a larger volume of water than the surface from where the oil was dispersed. They can also delay the formation of persistent oil-in-water emulsions. However, laboratory experiments showed that dispersants increased toxic hydrocarbon levels in fish by a factor of up to 100 and may kill fish eggs. Dispersed oil droplets infiltrate into deeper water and can lethally contaminate coral. Research indicates that some dispersants are toxic to corals. A 2012 study found that Corexit dispersant had increased the toxicity of oil by up to 52 times.

- Watch and wait: in some cases, natural attenuation of oil may be most appropriate, due to the invasive nature of facilitated methods of remediation, particularly in ecologically sensitive areas such as wetlands.

- Dredging: for oils dispersed with detergents and other oils denser than water.

- Skimming: Requires calm waters at all times during the process.

- Solidifying: Solidifiers are composed of tiny, floating, dry ice pellets, and hydrophobic polymers that both adsorb and absorb. They clean up oil spills by changing the physical state of spilled oil from liquid to a solid, semi-solid or a rubber-like material that floats on water. Solidifiers are insoluble in water, therefore the removal of the solidified oil is easy and the oil will not leach out. Solidifiers have been proven to be relatively non-toxic to aquatic and wild life and have been proven to suppress harmful vapors commonly associated with hydrocarbons such as Benzene, Xylene, Methyl Ethyl, Acetone and Naphtha. The reaction time for solidification of oil is controlled by the surface area or size of the polymer or dry pellets as well as the viscosity and thickness of the oil layer. Some solidifier product manufactures claim the solidified oil can be thawed and used if frozen with dry ice or disposed of in landfills, recycled as an additive in asphalt or rubber products, or burned as a low ash fuel. A solidifier called C.I.Agent (manufactured by C.I.Agent Solutions of Louisville, Kentucky) is being used by BP in granular form, as well as in Marine and Sheen Booms at Dauphin Island and Fort Morgan, Alabama, to aid in the Deepwater Horizon oil spill cleanup.

- Vacuum and centrifuge: oil can be sucked up along with the water, and then a centrifuge can be used to separate the oil from the water - allowing a tanker to be filled with near pure oil. Usually, the water is returned to the sea, making the process more efficient, but allowing small amounts of oil to go back as well. This issue has hampered the use of centrifuges due to a United States regulation limiting the amount of oil in water returned to the sea.

- Beach Raking: coagulated oil that is left on the beach can be picked up by machinery.

Bags of oily waste from the Exxon Valdez oil spill

Equipment used includes:

- Booms: large floating barriers that round up oil and lift the oil off the water

- Skimmers: skim the oil

- Sorbents: large absorbents that absorb oil

- Chemical and biological agents: helps to break down the oil

- Vacuums: remove oil from beaches and water surface

- Shovels and other road equipment: typically used to clean up oil on beaches

Prevention

- Secondary containment - methods to prevent releases of oil or hydrocarbons into environment.

- Oil Spill Prevention Containment and Countermeasures (SPCC) program by the United States Environmental Protection Agency.

- Double-hulling - build double hulls into vessels, which reduces the risk and severity of a spill in case of a collision or grounding. Existing single-hull vessels can also be rebuilt to have a double hull.

- Thick-hulled railroad transport tanks

Spill response procedures should include elements such as;

- A listing of appropriate protective clothing, safety equipment, and cleanup materials required

for spill cleanup (gloves, respirators, etc.) and an explanation of their proper use;

- Appropriate evacuation zones and procedures;

- Availability of fire suppression equipment;

- Disposal containers for spill cleanup materials; and

- The first aid procedures that might be required.

Environmental Sensitivity Index (ESI) Mapping

Environmental Sensitivity Index (ESI) maps are used to identify sensitive shoreline resources prior to an oil spill event in order to set priorities for protection and plan cleanup strategies. By planning spill response ahead of time, the impact on the environment can be minimized or prevented. Environmental sensitivity index maps are basically made up of information within the following three categories: shoreline type, and biological and human-use resources.

Shoreline Type

Shoreline type is classified by rank depending on how easy the target site would be to clean up, how long the oil would persist, and how sensitive the shoreline is. The floating oil slicks put the shoreline at particular risk when they eventually come ashore, covering the substrate with oil. The differing substrates between shoreline types vary in their response to oiling, and influence the type of cleanup that will be required to effectively decontaminate the shoreline. In 1995, the US National Oceanic and Atmospheric Administration extended ESI maps to lakes, rivers, and estuary shoreline types. The exposure the shoreline has to wave energy and tides, substrate type, and slope of the shoreline are also taken into account—in addition to biological productivity and sensitivity. The productivity of the shoreline habitat is also taken into account when determining ESI ranking. Mangroves and marshes tend to have higher ESI rankings due to the potentially long-lasting and damaging effects of both the oil contamination and cleanup actions. Impermeable and exposed surfaces with high wave action are ranked lower due to the reflecting waves keeping oil from coming onshore, and the speed at which natural processes will remove the oil.

Biological Resources

Habitats of plants and animals that may be at risk from oil spills are referred to as "elements" and are divided by functional group. Further classification divides each element into species groups with similar life histories and behaviors relative to their vulnerability to oil spills. There are eight element groups: Birds, Reptiles, Amphibians, Fish, Invertebrates, Habitats and Plants, Wetlands, and Marine Mammals and Terrestrial Mammals. Element groups are further divided into sub-groups, for example, the 'marine mammals' element group is divided into dolphins, manatees, pinnipeds (seals, sea lions & walruses), polar bears, sea otters and whales. Problems taken into consideration when ranking biological resources include the observance of a large number of individuals in a small area, whether special life stages occur ashore (nesting or molting), and whether there are species present that are threatened, endangered or rare.

Human-use Resources

Human use resources are divided into four major classifications; archaeological importance or cultural resource site, high-use recreational areas or shoreline access points, important protected management areas, or resource origins. Some examples include airports, diving sites, popular beach sites, marinas, natural reserves or marine sanctuaries.

Estimating The Volume of A Spill

By observing the thickness of the film of oil and its appearance on the surface of the water, it is possible to estimate the quantity of oil spilled. If the surface area of the spill is also known, the total volume of the oil can be calculated.

Appearance	Film thickness			Quantity spread	
	inches	mm	nm	gal/sq mi	L/ha
Barely visible	0.0000015	0.0000380	38	25	0.370
Silvery sheen	0.0000030	0.0000760	76	50	0.730
First trace of color	0.0000060	0.0001500	150	100	1.500
Bright bands of color	0.0000120	0.0003000	300	200	2.900
Colors begin to dull	0.00004	0.0010000	1000	666	9.700
Colors are much darker	0.0000800	0.0020000	2000	1332	19.500

Oil spill model systems are used by industry and government to assist in planning and emergency decision making. Of critical importance for the skill of the oil spill model prediction is the adequate description of the wind and current fields. There is a worldwide oil spill modelling (WOSM) program. Tracking the scope of an oil spill may also involve verifying that hydrocarbons collected during an ongoing spill are derived from the active spill or some other source. This can involve sophisticated analytical chemistry focused on finger printing an oil source based on the complex mixture of substances present. Largely, these will be various hydrocarbons, among the most useful being polyaromatic hydrocarbons. In addition, both oxygen and nitrogen heterocyclic hydrocarbons, such as parent and alkyl homologues of carbazole, quinoline, and pyridine, are present in many crude oils. As a result, these compounds have great potential to supplement the existing suite of hydrocarbons targets to fine-tune source tracking of petroleum spills. Such analysis can also be used to follow weathering and degradation of crude spills.

Oil–Water Separator

An oil water separator (OWS) is a piece of equipment used to separate oil and water mixtures into their separate components. There are many different types of oil-water separator. Each has different oil separation capability and are used in different industries. Oil water separators are designed and selected after consideration of oil separation performance parameters and life cycle cost considerations. "Oil" can be taken to mean mineral, vegetable and animal oils, and the many different hydrocarbons.

Oil water separators can be designed to treat a variety of contaminants in water including free floating oil, emulsified oil, dissolved oil and suspended solids. Not all oil separator types are capable of separating all contaminants. The most common performance parameters considered are:

- Oil droplet size (in the feed to the separator)

- Oil density

- Water viscosity (temperature)

- Discharge water quality desired

- Feed oil concentration and the range of oil concentrations likely

- Feed oil water flow (daily and peak hourly)

API oil–water Separator

An API oil–water separator is a device designed to separate gross amounts of oil and suspended solids from the wastewater effluents of oil refineries, petrochemical plants, chemical plants, natural gas processing plants and other industrial sources. The name is derived from the fact that such separators are designed according to API Publication 421, February 1990, published by the American Petroleum Institute. These separators can be used to separate large oil droplets, typically greater than 150 micron.

Oily Water Separator (Marine)

Marine oily water separator

The purpose of shipboard oily water separator (OWS) is to separate oil and other contaminants that could be harmful for the oceans. They are most commonly found on board ships where they are used to separate oil from oily waste water such as bilge water before the waste water is discharged into the environment. These discharges of waste water must comply with the requirements laid out in Marpol 73/78.

Bilge water is a near-unavoidable product of shipboard operations. Oil leaks from running machinery, such as diesel generators, air compressors, and the main propulsion engine. Modern OWSs have alarms and automatic closure devices which are activated when the oil storage capacity of the oil water separator has been reached.

Gravity Plate Separator

A gravity plate separator contains a series of plates through which the contaminated water flows. The objective of the design is to allow oil droplets in the water to coalesce on the underside of the plate eventually forming larger oil droplets which floats off the plates and accumulates at the top of the chamber. The oil accumulating at the top is then transferred with some en-trained water to a waste oil tank. This type of oily water separator is very common for many industrial applications as well as in ships but it has some flaws that decrease efficiency. Oil particles that are sixty micrometers in size or smaller do not get separated. Also the presence of chemicals and surfactants in the water greatly reduce oil droplet coalescence, impeding the separation effect The variety of oily wastes in bilge water can limit removal efficiency especially when very dense and highly viscous oils such as bunker oil are present. Plates must be replaced when fouled, which increases the costs of operation.

Centrifugal Oily Water Separator

Centrifuge oily water separator

A centrifugal water–oil separator, *centrifugal oil–water separator* or *centrifugal liquid–liquid separator* is a device designed to separate oil and water by centrifugation. It generally contains a cylindrical container that rotates inside a larger stationary container. The denser liquid, usually water, accumulates at the periphery of the rotating container and is collected from the side of the device, whereas the less dense liquid, usually oil, accumulates at the rotation axis and is collected from the center. Centrifugal oil–water separators are used for waste water processing and for cleanup of oil spills on sea or on lake. Centrifugal oil–water separators are also used for filtering diesel and lubricating oils by removing the waste particles and impurity from them.

Hydrocyclone Oily Water Separator

An oil water separation hydrocyclone is a device designed to separate oil from water by the use of a strong vortex. These separators are passive (no moving parts) and resemble long tapered pipes. They typically contains an inlet section, long tapered section and a long outlet section. In operation the strong vortex is created when the oily water is injected tangentially into the inlet end of the separator. This creates a centrifugal force, that accelerates as it moves down the tapered cone. The centripetal and centrifugal forces separate the heavier water component to the outside of the vortex while the lighter oil droplets are forced to the centre. The separated oils are removed through an orifice at the inlet end of the cone and treated water is discharged through the opposite end. The centrifugal forces generated inside the vortex of the better de-oiling hydrocyclone separators are of the order of 1,000 times the force of gravity. This is why smaller emulsified oil droplets as low as 15 microns can be removed.

Oil removal hydrocyclones, or de-oiling hydrocyclones, are very different in geometry, design and operation compared to the more common solid removal hydrocyclones. When correctly designed and operated oil removal hydrocyclones Hydrocyclones are very useful for removing both large oil droplets and smaller emulsified oil droplets in a broad range of applications across many industries. The technology has been successfully applied to treat oily water produced in the mining industry, meat processing, dairy manufacturing, petrochemical, oil refining, oil marketing and oil production operations.

Electrochemical

Wastewater purification of oils and contaminates by electrochemical emulsification is actively in research and development. Electrochemical emulsification involves the generation of electrolytic bubbles that attract pollutants such as sludge and carry them to the top of the treatment chamber. Once at the top of the treatment chamber the oil and other pollutants are transferred to a waste oil tank.

Downhole Oil–Water Separation

Downhole oil–water separation (DOWS) technology is an emerging technology that separates oil and gas from produced water at the bottom of the well, and re-injects most of the produced water into another formation which is usually deeper than the producing formation, while the oil and gas rich stream is pumped to the surface. DOWS effectively removes solids from the disposal fluid and thus avoids injectivity impairment caused by solids plugging. Simultaneous injection using DOWS minimizes the opportunity for the contamination of underground sources of drinking water (USDWs) through leaks in tubing and casing during the injection process.

Bioremediation

Bioremediation is the use of microorganisms to treat contaminated water. A carefully managed environment is needed for the microorganisms which includes nutrients and hydrocarbons such as oil or other contaminates, and oxygen.

In pilot scale studies, bio-remediation was used as one stage in a multi-stage purification process involving a plate separator to remove the majority of the contaminants and was able to treat pollutants at very low concentrations including organic contaminates such as glycerol, solvents, jet fuel, detergents, and phosphates. After treatment of contaminated water, carbon dioxide, water and an organic sludge were the only residual products.

API Oil–Water Separator

An API oil–water separator is a device designed to separate gross amounts of oil and suspended solids from the wastewater effluents of oil refineries, petrochemical plants, chemical plants, natural gas processing plants and other industrial oily water sources. The name is derived from the fact that such separators are designed according to standards published by the American Petroleum Institute (API).

Description of The Design and Operation

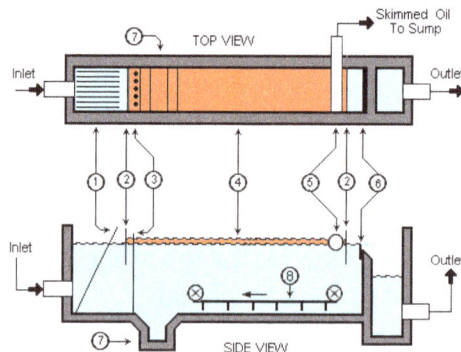

1 Trash trap (inclined rods)
2 Oil retention baffles
3 Flow distributors (vertical rods)
4 Oil layer
5 Slotted pipe skimmer
6 Adjustable overflow weir
7 Sludge sump
8 Chain and flight scraper

A typical gravimetric API separator

The API separator is a gravity separation device designed using Stokes' law principles that define the rise velocity of oil droplets based on their density, size and water properties. The design of the separator is based on the specific gravity difference between the oil and the wastewater because that difference is much smaller than the specific gravity difference between the suspended solids and water. Based on that design criterion, most of the suspended solids will settle to the bottom of the separator as a sediment layer, the oil will rise to top of the separator, and the wastewater will be the middle layer between the oil on top and the solids on the bottom. The API Design Standards, when correctly applied, make adjustments to the geometry, design and size of the separator beyond simple Stokes Law principles. This includes allowances for water flow entrance and exit turbulence losses as well as other factors.

Typically in operation of API separators the oil layer, which may contain en-trained water and attached suspended solids, is continually skimmed off. This removed oily layer may be re-processing to recover valuable products, or disposed of. The heavier bottom sediment layer is removed by a chain and flight scraper (or similar device) and a sludge pump.

Design Limitations

API design separators, and similar gravity tanks, are not intended to be effective when any of the following conditions apply to the feed conditions:

- Mean Oil droplets size in the feed is less than 150 micron

- Oil density is greater than 925 kg/m3

- Suspended solids are adhering to the oil meaning the 'effective' oil density is greater than 925 kg/m3

- Water temperature less than 5 oC

- There are high levels of dissolved hydrocarbons

Further Treatment of API Water Discharges

Because of performance limitations the water discharged from API type separators usually requires several further processing stages before the treated water can be discharged or reused. Further water treatment is designed to remove oil droplets smaller than 150 micron, dissolved materials and hydrocarbons, heavier oils or other contaminants not removed by the API. Secondary treatment technologies include dissolved air flotation (DAF), Anaerobic and Aerobic biological treatment, Parallel Plate Sseparators, Hydrocyclone, Walnut Shell Filters and Media filters.

Plate separators, or Coalescing Plate Separators are similar to API separators, in that they are based on Stokes Law principles, but include inclined plate assemblies (also known as parallel packs). The underside of each parallel plate provides more surface for suspended oil droplets to coalesce into larger globules. Coalescing plate separators may not be effective in situation where water chemicals or suspended solids restrict or prevent oil droplets coalesce. In operation it is intended that sediment will slide down the topside of each parallel plate, however in many practical situations the sediment can adhear to the plates requiring periodic removal and cleaning. Such separators still depend upon the specific gravity between the suspended oil and the water. However, the parallel plates can enhance the degree of oil-water separation for oil droplets above 50 micron in size. Alternatively parallel plate separators are added to the design of API Separators and require less space than a conventional API separator to achieve a similar degree of separation.

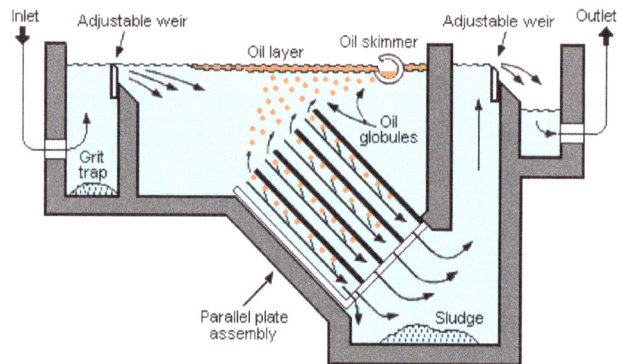

A typical parallel plate separator

History

The API separator was developed by the API and the Rex Chain Belt Company (now Siemens Water). The first API separator was installed in 1933 at the Atlantic Refining Company (ARCO) refinery in Philadelphia. Since that time, virtually all of the refineries worldwide have installed API separators as a first primary stage of their oily wastewater treatment plants. The majority of those refineries installed the API separators using the original design based on the specific gravity difference between oil and water. However, many refineries now use plastic parallel plate packing to enhance the gravity separation.

Other oil–Water Separation Applications

There are other applications requiring oil-water separation. For example:

- Oily water separators (OWS) for separating oil from the bilge water accumulated in ships as required by the international MARPOL Convention.

- Oil and water separators are commonly used in electrical substations. The transformers found in substations use a large amount of oil for cooling purposes. Moats are constructed surrounding unenclosed substations to catch any leaked oil, but these will also catch rainwater. Oil and water separators therefore provide a quicker and easier cleanup of an oil leak.

Centrifugal Water–Oil Separator

A centrifugal water–oil separator, centrifugal oil–water separator or centrifugal liquid–liquid separator is a device designed to separate oil and water by centrifugation. It generally contains a cylindrical container that rotates inside a larger stationary container. The denser liquid, usually water, accumulates at the periphery of the rotating container and is collected from the side of the device, whereas the less dense liquid, usually oil, accumulates at the rotation axis and is collected from the centre.

Centrifugal oil–water separators are used for waste water processing and for cleanup of oil spills on sea or on lake.

Centrifugal oil–water separators are also used for filtering diesel and lubricating oils by removing the waste particles and impurity from them.

Mechanism

A mix of oil and water is pumped constantly into a cone-shaped separating apparatus at an angle, which creates a spinning vortex. The filtration is a result of the force balance that occurs on fluids in a vortex. High-density liquids will move to the outside, along with any contaminant, displacing the lower-density liquids to the inside (center of rotation). Water, being the more dense liquid, sits on the outside and is removed through a discharge outlet. Any segregated oil can now safely be recovered through a suction orifice at the center. The process will continue to function in this fashion as long as sufficient oil is added to maintain coverage of the suction orifice.

Comparison to Other Types of Separators

There are other types of separators that use gravitational forces to separate mixtures, but these other types of forces are not as strong as the centrifugal force in the centrifugal separator. Other types of separators are coalescing plate pack separators and petrol interceptor separators.

Coalescing plate pack separators work very differently from centrifugal separators. With the plate packs, water is fed into the separator through gravity through the inlet pipe, then the mixture is spread evenly through the separation chamber where the coalescing plate packs are. In the plate packs the oil will rise because of their buoyancy and coalesce on the underside of the plates and form globules of oil that rise to the surface. From there the waste oil globules go into the clean water chamber and are discharged through the lower portion of the separator.

With the interceptors the dirty water mixture enters the first tank of the interceptor where that tanks builds up hydrocarbons and other hazardous material in a layer. Then comes the second and third tanks of the interceptor that are all connected through pipes called "dip pipes" which don't allow the hydrocarbons and other contaminants to pass through. The same process that happened in the first tank will happen in the second and third tanks just with less and less contaminants.

In a centrifugal oil and water separator, the force of gravity is one-thousand times greater that of the coalescing plate pack separator or the petrol interceptor, so the separation is much greater. Not only is the force of separation greater, but there are fewer working parts so maintenance is much easier and cheaper.

More advantages of the centrifugal oil and water separator include compact equipment size, versatility, ease of use, low cost, and high performance.

Disadvantages

Centrifugal oil and water separators do have their disadvantages. One known disadvantage of these separators is that they tend to have low powered suction. For example, when the pump end is dry and the impeller is rotating at high speeds, it is simply not powerful enough to lift the oily water mixture into the separator. For this reason, these separators must always be primed before use.

Design and Features

Materials and Metals Used

The bowl of the separator is generally made up of stainless steel, brass, and bronze. The structural parts are usual where the stainless steel is and the parts on the inside that come into contact with mixtures are either made of bronze or brass depending on the mixture going through the centrifugal separator. The housing and gearbox of the separator is made of aluminum or stainless steel and cast iron. The gearbox in particular is cast iron with stainless steel coating it.

Transmission

The vortex of the separators is generated through a power source and the power source in use for the centrifugal separator is an electrical motor and the pieces of its transmission include a clutch and worm gear, a flexible coupling worm gear, and a set of pulleys and fat belt.

Suction and Discharging of The Liquids

The suction process of the separator takes place as stationary feed pipe suctions the mixture liquid into the cone-shaped pipe that feeds into the vortex. After the filtration of the liquids occur the unwanted contaminants are discharged either through an overflow or through one or two centripetal pumps.

Oily water separator

Hydrocyclone

A hydrocyclone (often referred to by the shortened form cyclone) is a device to classify, separate or sort particles in a liquid suspension based on the ratio of their centripetal force to fluid resistance. This ratio is high for dense (where separation by density is required) and coarse (where separation by size is required) particles, and low for light and fine particles. Hydrocyclones also find application in the separation of liquids of different densities.

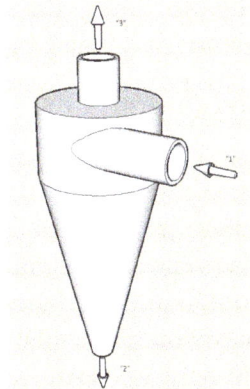

Diagram of a hydrocyclone:
1. the liquid-solid mixture enters,
2. heavy solids leave,
3. cleaned liquid leaves.

A different description: A hydrocyclone is a mechanical device designed to reduce or increase the concentration of a dispersed phase, solid, liquid or gas of different density, by means of centripetal forces or centrifugal forces within a vortex.

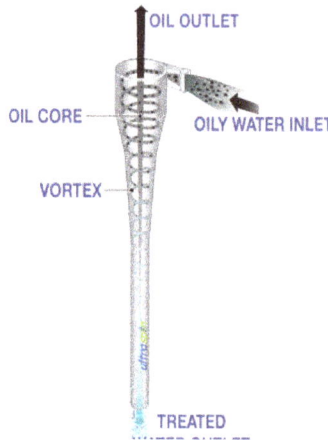

Centrifugal forces in a hydrocyclone for oily water separation

The mixture is injected into the hydrocyclone in such a way as to create the vortex and, depending upon the relative densities of the two phases, the centrifugal acceleration will cause the dispersed phase to move away from or towards the central core of the vortex.

A hydrocyclone will normally have a cylindrical section at the top where liquid is being fed tangentially, and a conical base. The angle, and hence length of the conical section, plays a role in determining operating characteristics.

Design

A hydrocyclone is a classifier that has two exits on the axis: one on the bottom (*under-flow* or *reject*) and one at the top (*overflow* or *accept*). The underflow is generally the denser or coarser fraction, while the overflow is the lighter or finer fraction. It has no moving parts and its operation depends two major parameters:

- the characteristics of the feed stream.

- the geometry of the cyclone.

The characteristics of the feed stream include size distribution of solids in the feed stream, pulp density (percent solids in the slurry), pulp viscosity and the inlet pressure for solid/liquid separation. In liquid/liquid feed streams, for example in oily water, the main feed characteristics are based on oil droplet size and distribution, oil density, water density, oil concentration, viscosity and temperature.

The geometry of the cyclone involves-inlet shape and area, cyclone dimensions (cone angle, length of cylindrical section and total length of the cyclone) and inlet, vortex and apex diameters.

Internally, inertia is countered by the resistance of the liquid, with the effect that larger or denser particles are transported to the wall for eventual exit at the underflow side with a limited amount of liquid, while the finer, or less dense particles, remain in the liquid and exit at the overflow side through a tube extending slightly into the body of the cyclone at the center.

Forward hydrocyclones remove particles that are denser than the surrounding fluid, while reverse hydrocyclones remove particles that are less dense than the surrounding fluid. In a reverse hydrocyclone the overflow is at the apex and the underflow at the base. There are also parallel-flow hydrocyclones where both the *accept* and *reject* are removed at the apex. Parallel-flow hydrocyclones remove particles that are lighter than the surrounding fluid.

Hydrocyclones can be made of metal (mostly steel), ceramic or plastic (such as polyurethane, polypropylene, or other types). Metal or ceramic hydrocyclones are used for situations requiring more strength, or durability in terms of heat or pressure. When there is an occurrence of much abrasion (such as occurs with sand particles) polyurethane performs better than metals or ceramics. Metal lined with polyurethane is used in cases of combined abrasion and high pressure.

In a suspension of particles with the same density, a relatively sharp cut can be made. The size at which the particles separate is a function of cyclone diameter, exit dimensions, feed pressure and the relative characteristics of the particles and the liquid. Efficiency of separation is a function of the solids' concentration: the higher the concentration, the lower the efficiency of separation. There is also a significant difference in suspension density between the base exit (fines) and the apex exit, where there is little liquid flow.

If the size range of the particles is limited, but there are differences in density between types of particles, the denser particles will exit preferentially at the apex. The device is therefore a means of selective concentration of, for example, minerals.

This device is also related to the centrifuge; both of them are intended to separate heavies and lights in liquid by application of centrifugal force. Centrifuges generate the separation force by rotation of the entire equipment; hydrocyclones utilise centrifugal forces from the movement of the fluids to achieve separation.

Design Equations

For a Bradley hydrocyclone, the dimensions of each part are always in the same proportions. The pressure drop should be between 20 psi and 60 psi.

Di is the inlet diameter

Do is the overflow diameter

Du is the underflow diameter

Dc is the chamber diameter

L is the height of the hydrocyclone

l is the height of the cylinder part of hydrocyclone

ℓ is the height of the vortex tube of hydrocyclone

Le is the length of the outlet tube

Theta is the angle at the base of the hydrocyclone

Di/Dc = 1/7

Do/Dc = 1/5

Du/Dc = -

L/Dc = -

l/Dc = 1/2

ℓ/Dc = 1/3

Theta = 9 degrees

Uses

A hydrocyclone is most often used to separate "heavies" from a liquid mixture origi-nating at a centrifugal pump or some other continuous source of pressurized liquid. A hydrocyclone is most likely to be the right choice for processes where "lights" are the greater part of the mixture and where the "heavies" settle fairly easily.

Generally, hydrocyclones are used in continuous flow systems so that the instantaneous liquid inflow to the hydrocyclone is equal to the total instantaneous outflow of "lights" plus "heavies". In cases where "heavies" are a very small part of the whole liquid, it is sometimes advantageous to accumulate them in the bottom of the hydrocyclone for batchwise removal.

Applications include:

- In the potato, cassava, wheat and corn starch industry for both concentration and washing of the crude starch milk. Hydrocyclone replace separators as a more price efficient separation technique.

- In pulp and paper mills to remove sand, staples, plastic particles and other con-taminants.

- In the drilling industry to separate sand from the expensive clay that is used for lubrication during the drilling.

- In oil industry to separate oil from water or *vice versa*.

- In metal working to separate metal particles from cooling liquid.

- In French fries and potato chips plants for in-line starch recovery from cutting water and from waste water.

- In mineral processing, hydrocyclones are used extensively both to classify particles for recirculation in grinding circuits and to differentiate between the economic mineral and gangue.

- To remove sand and silt particles from irrigation water for drip irrigation purposes.

References

- "Dispersant makes oil 52 times more toxic - Technology & science - Science - LiveScience - NBC News". msnbc.com. Retrieved 20 April 2015.

- "Separation and Purification Technology" (PDF). www.atmc.umassd.edu. Archived from the original (PDF) on 2015-09-24. Retrieved 2015-06-03.

- "Lingering Lessons of the Exxon Valdez Oil Spill". Commondreams.org. 2004-03-22. Archived from the original on June 13, 2010. Retrieved 2012-08-27.

- Campbell Robertson /Clifford Krauss (2 August 2010). "Gulf Spill Is the Largest of Its Kind, Scientists Say". The New York Times. New York Times. Retrieved 2 August 2010.

- National Oceanic and Atmospheric Administration, Office of Response and Restoration, Emergency Response Division, Incident News: Arabian Gulf Spills, updated 18 May 2010.

- Consumer Energy Report (20 June 2010). "Internal Documents: BP Estimates Oil Spill Rate up to 100,000 Barrels Per Day". Consumer Energy Report. Retrieved 20 June 2010.

- "Spill Response - Dispersants Kill Fish Eggs". journal Environmental Toxicology and Chemistry. Retrieved 2010-05-21.

- Fountain, Henry (2010-06-24). "Advances in Oil Spill Cleanup Lag Since Valdez". New York Times. Retrieved 2010-07-05.

Conservartion and Resource Reclamation Projects

Conservation and resource reclamation is a developing field of study; the following chapter provides an outline, and also helps the reader to understand the subject of conservation. Some of the projects explained in this chapter are reclaimed water techniques, eco industrial park and industrial ecology.

Reclaimed Water

Reclaimed water or recycled water, is former wastewater (sewage) that is treated to remove solids and impurities, and used in sustainable landscaping irrigation, to recharge groundwater aquifers, to meet commercial and industrial water needs, and for drinking. The purpose of these processes is water conservation and sustainability, rather than discharging the treated water to surface waters such as rivers and oceans. In some cases, recycled water can be used for streamflow augmentation to benefit ecosystems and improve aesthetics. One example of this is along Calera Creek in the City of Pacifica, CA.

Samples of different types of (waste)water, starting with raw sewage then plant effluent and finally reclaimed water (after several treatment steps)

The definition of reclaimed water, as defined by Levine and Asano, is "The end product of wastewater reclamation that meets water quality requirements for biodegradable materials, suspended matter and pathogens." Simply stated, reclaimed water is water that is used more than one time before it passes back into the natural water cycle. Scientifically-proven advances in water technology allow communities to reuse water for many different purposes, including industrial, irrigation, and drinking. The water is treated differently depending upon the source and use of the water and how it gets delivered.

Cycled repeatedly through the planetary hydrosphere, all water on Earth is recycled water, but the terms "recycled water" or "reclaimed water" typically mean wastewater sent from a home or business through a pipeline system to a treatment facility, where it is treated to a level consistent with its intended use. The water is then routed directly to a recycled water system for uses such as irrigation or industrial cooling.

There are examples of communities that have safely used recycled water for many years. Los Angeles County's sanitation districts have provided treated wastewater for landscape irrigation in parks and golf courses since 1929. The first reclaimed water facility in California was built at San Francisco's Golden Gate Park in 1932. The Irvine Ranch Water District (IRWD) was the first water district in California to receive an unrestricted use permit from the state for its recycled water; such a permit means that water can be used for any purpose except drinking. IRWD maintains one of the largest recycled water systems in the nation with more than 400 miles serving more than 4,500 metered connections. The Irvine Ranch Water District and Orange County Water District in Southern California are established leaders in recycled water. Further, the Orange County Water District, located in Orange County, and in other locations throughout the world such as Singapore, water is given more advanced treatments and is used indirectly for drinking.

In spite of quite simple methods that incorporate the principles of water-sensitive urban design (WSUD) for easy recovery of stormwater runoff, there remains a common perception that reclaimed water must involve sophisticated and technically complex treatment systems, attempting to recover the most complex and degraded types of sewage. As this effort is driven by sustainability factors, this type of implementation should inherently be associated with point source solutions, where it is most economical to achieve the expected outcomes. Harvesting of stormwater or rainwater can be an extremely simple to comparatively complex, as well as energy and chemical intensive, recovery of more contaminated sewage.

Terminology

Effluent storage tank from where treated effluent (after constructed wetland) is pumped away for irrigation, Haran-Al-Awamied, Syria

There is no one-size-fits-all solution to water reuse, but there are many safe and scientifically-proven options that allow communities to sustain their local water supplies. Below are terms scientists and water experts use to describe some of these reclaimed water options:

Reused water is water used more than once or recycled.

Potable water is drinking water.

Potable reuse refers to reused water you can drink.

Nonpotable reuse refers to reused water that is not used for drinking, but is safe to use for irrigation or industrial purposes.

De facto, unacknowledged or unplanned potable reuse occurs when water intakes draw raw water supplies downstream from discharges of treated effluent from wastewater treatment plants/water reclamation facilities or resource recovery facilities. For example, if you are downstream of a community, that community's used water (run-off and treated wastewater) gets put back into river or stream and is delivered downstream to your community and becomes part of your drinking water supply.

Planned potable reuse is publicly acknowledged as an intentional project to recycle water for drinking water. It can be either direct or indirect. It commonly involves a more formal public process and public consultation program than is observed with de facto or unacknowledged reuse.

How potable reused water is delivered determines if it is called Indirect Potable Reuse or Direct Potable Reuse.

- Indirect potable reuse means the water is delivered to you indirectly. After it is purified, the reused water blends with other supplies and/or sits a while in some sort of storage, man-made or natural, before it gets delivered to a pipeline that leads to a water treatment plant or distribution system. That storage could be a groundwater basin or a surface water reservoir.

- Direct potable reuse means the reused water is put directly into pipelines that go to a water treatment plant or distribution system. Direct potable reuse may occur with or without "engineered storage" such as underground or above ground tanks.

Greywater uses the same waste as water reclamation with the exception of toilet water. Greywater can be used for irrigation, toilet flushing or other domestic uses.

Desalination is an energy-intensive process where salt and other minerals are removed from sea water to produce potable water for drinking and irrigation, typically through membrane filtration (reverse-osmosis), and steam-distillation.

Maximum water recovery - To determine maximum water recovery there are various techniques that have been developed by researchers; for maximum water reuse/reclamation/recovery strategies such as water pinch analysis. The techniques help a user to target the minimum freshwater consumption and wastewater target. It also helps in designing the network that achieves the target. This provides a benchmark to be used by users in improving their water systems.

Applications

Most of the uses of water reclamation are non potable uses such as: Washing Cars, flushing toilets, cooling water for power plants, concrete mixing, artificial lakes, irrigation for golf courses and public parks, and for hydraulic fracturing. Where applicable, systems run a dual piping system to keep the recycled water separate from the potable water.

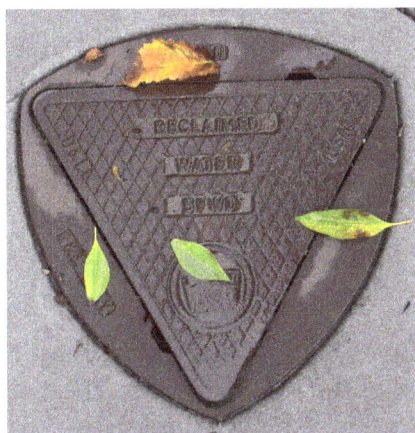

Cover for reclaimed water valve, San Francisco Water District

New technologies for recycling allow the water to be used for fracking purposes and can save an estimated 4 - 7 million or more gallons per well.

Potable Uses

Some water agencies reuse highly treated effluent from municipal wastewater or resource recovery plants as a reliable, drought proof source of drinking water. By using advanced purification processes, they produce water that meets all applicable drinking water standards. System reliability and frequent monitoring and testing are imperative to them meeting stringent controls.

The water needs of a community, water sources, public health regulations, costs, and the types of water infrastructure in place, such as distribution systems, man-made reservoirs, or natural groundwater basins, determine if and how reclaimed water can be part of the drinking water supply. Communities in El Paso, Texas and Orange County, California, for example, reuse water to replenish groundwater basins. Others, such as the Upper Occoquan Service Authority in Virginia, put it into surface water reservoirs. In these instances the reclaimed water is blended with other water supplies and/or sits in storage for a certain amount of time before it is drawn out and gets treated again at a water treatment or distribution system. In some Texas communities, the reused water is put directly into pipelines that go to a water treatment plant or distribution system. In Singapore reclaimed water is called NEWater and is bottled directly from an advanced water purification facility for educational and celebratory purposes. Though most of the reused water is used for high-tech industry in Singapore, a small amount is returned to reservoirs for drinking water.

A 2012 study conducted by the National Research Council in the United States of America found that the risk of exposure to certain microbial and chemical contaminants from drinking reclaimed water does not appear to be any higher than the risk experienced in at least some current drinking water treatment systems, and may be orders of magnitude lower. This report recommends adjustments to the federal regulatory framework that could enhance public health protection for both planned and unplanned (or de facto) reuse and increase public confidence in water reuse.

Modern technologies such as reverse osmosis and ultraviolet disinfection are commonly used when reclaimed water will be mixed with the drinking water supply. An experiment by the University of New South Wales reportedly showed a reverse osmosis system removed ethinylestradiol and paracetamol from the wastewater, even at 1000 times the expected concentration.

Aboard the International Space Station, astronauts have been able to drink recycled urine due to the introduction of the ECLSS system. The system costs $250 million and has been working since May 2009. The system recycles wastewater and urine back into potable water used for drinking, food preparation, and oxygen generation. This cuts back on the need for resupplying the space station so often.

Indirect Potable Reuse

Some municipalities are using and others are investigating Indirect Potable Reuse (IPR) of reclaimed water. For example, reclaimed water may be pumped into (subsurface recharge) or percolated down to (surface recharge) groundwater aquifers, pumped out, treated again, and finally used as drinking water. This technique may also be referred to as *groundwater recharging*. This includes slow processes of further multiple purification steps via the layers of earth/sand (absorption) and microflora in the soil (biodegradation).

Direct Potable Reuse

In a Direct Potable Reuse (DPR) scheme, water is put directly into pipelines that go to a water treatment plant or distribution system. Direct potable reuse may occur with or without "engineered storage" such as underground or above ground tanks. Communities in Texas have implemented DPR projects, and the state of California is studying the feasibility of developing DPR regulations.

Unplanned Potable Reuse

Water reuse occurs in various ways throughout the world. It happens daily on rivers and other water bodies everywhere. If you live in a community downstream of another, chances are you are reusing its water and likewise communities downstream of you are most likely reusing your water. Unplanned Indirect Potable Use has existed even before the introduction of reclaimed water. Many cities already use water from rivers that contain effluent discharged from upstream sewage treatment plants. There are many large towns on the River Thames upstream of London (Oxford, Reading, Swindon, Bracknell) that discharge their treated sewage ("non-potable water") into the river, which is used to supply London with water downstream.

This phenomenon is also observed in the United States, where the Mississippi River serves as both the destination of sewage treatment plant effluent and the source of potable water. Research conducted in the 1960s by the London Metropolitan Water Board demonstrated that the maximum extent of recycling water is about 11 times before the taste of water induces nausea in sensitive individuals. This is caused by the buildup of inorganic ions such as Cl^-, SO_4^{2-}, K^+ and Na^+, which are not removed by conventional sewage treatment.

Space Travel

Wastewater reclamation can be especially important in relation to human spaceflight.

- In 1998, NASA announced it had built a human waste reclamation bioreactor designed for use in the International Space Station and a manned Mars mission. Human urine and feces are input into one end of the reactor and pure oxygen, pure water, and compost (humanure) are output from the other end. The soil could be used for growing vegetables, and the bioreactor also produces electricity.

Design Considerations

Distribution

Nonpotable reclaimed water is often distributed with a dual piping network that keeps reclaimed water pipes completely separate from potable water pipes. In the United States and some other countries, nonpotable reclaimed water is distributed in lavender (light purple) pipes to distinguish it from potable water. The use of the color purple for pipes carrying recycled water was pioneered by the Irvine Ranch Water District in Irvine, California.

In many cities using reclaimed water, it is now in such demand that consumers are only allowed to use it on assigned days. Some cities that previously offered unlimited reclaimed water at a flat rate are now beginning to charge citizens by the amount they use.

Reclamation Processes

Wastewater must pass through numerous systems before being returned to the environment. Here is a partial listing from one particular plant system:

- Barscreens - Barscreens remove large solids that are sent into a grinder. All solids are then dumped into a sewer pipe at a Treatment Plant.

- Primary Settling Tanks - Readily settable and floatable solids are removed from the wastewater. These solids are skimmed from the top and bottom of the tanks and sent to the Treatment Plant where it'll be turned into fertilizer.

- Biological Treatment - The wastewater is cleaned through a biological treatment method that uses microorganisms, bacteria which digest the sludge and reduce the nutrient content. Air bubbles up to keep the organisms suspended and to supply oxygen to the aerobic bacteria so they can metabolize the food, convert it to energy, CO_2, and water, and reproduce more microorganisms. This helps to remove ammonia also through nitrification.

- Secondary Settling Tanks - The force of the flow slows down as sewage enters these tanks, allowing the microorganisms to settle to the bottom. As they settle, other small particles suspended in the water are picked up, leaving behind clear wastewater. Some of the microorganisms that settle to the bottom are returned to the system to be used again.

- Tertiary Treatment - Deep-bed, single-media, gravity sand filters receive water from the secondary basins and filter out the remaining solids. As this is the final process to remove solids, the water in these filters is almost completely clear.

- Chlorine Contact Tanks - Three chlorine contact tanks disinfect the water to decrease the risks associated with discharging wastewater containing human pathogens. This step protects the quality of the waters that receive the wastewater discharge.

One of two procedures are then followed according to the future disposal site:

1. Reclaimed Water Pump Station - The pump station distributes reclaimed water to users around the City. This may include golf courses, agricultural uses, cooling towers, or in land fills.

2. Water is passed through high level purification to be returned to the environment. Currently this means a reverse osmosis system.

Treatment Improvements

As world populations require both more clean water and better ways to dispose of wastewater, the demand for water reclamation will increase. Future success in water reuse will depend on whether this can be done without adverse effects on human health and the environment.

In the United States, reclaimed waste water is generally treated to secondary level when used for irrigation, but there are questions about the adequacy of that treatment. Some leading scientists in the main water society, AWWA, have long believed that secondary treatment is insufficient to protect people against pathogens, and recommend adding at least membrane filtration, reverse osmosis, ozonation, or other advanced treatments for irrigation water.

There have been recent advances in reverse osmosis in different countries, but have consistently produced very high quality water all the same. In Singapore, reclaimed water, also known as NEWater has become cleaner than the government issue tap water. Also, according to Bartels, the Bedok Demonstration Plant, which uses RO membranes, has successfully run for the past 3 years, producing high quality wastewater all the while.

Seepage of nitrogen and phosphorus into ground and surface water is also becoming a serious problem, and will probably lead to at least tertiary treatment of reclaimed water to remove nutrients in the future. Even using secondary treatment, water quality can be improved. Water quality can also be improved as it passes through the subsurface mixing zone where surface water and groundwater combine. Testing for pathogens using Polymerase Chain Reaction (PCR) instead of older culturing techniques, and changing the discredited fecal coloform "indicator organism" standard would be improvements.

In a large study treatment plants showed that they could significantly reduce the numbers of parasites in effluent, just by making adjustments to the currently used process. But, even using the best of current technology, risk of spreading drug resistance in the environment through wastewater effluent, would remain.

Some scientists have suggested that there need to be basic changes in treatment, such as using bacteria to degrade waste based on nitrogen (urine) and not just carbonaceous (fecal) waste, saying that this would greatly improve effectiveness of treatment. Currently designed plants do not deal well with contaminants in solution (e.g. pharmaceuticals). "Dewatering" solids is a major problem. Some wastes could be disposed of without mixing them with water to begin with. In an interesting innovation, solids (sludge) could be removed before entering digesters and burned into a gas that could be used to run engines.

Emerging disinfection technologies include ultrasound, pulse arc electrohydraulic discharge, and bank filtration. Another issue is concern about weakened mandates for pretreatment of industrial wastes before they are made part of the municipal waste stream. Some also believe that hospitals should treat their own wastes. The safety of drinking reclaimed water which has been given advanced treatment and blended with other waters, remains controversial.

In recent years, as hydraulic fracturing of oil and gas formations has become more and more common place, new technologies for water recycling have emerged. One such technology uses a combination of ozone and electrocoagulation. This process removes organics, hydrocarbons, spent polymers, chemical additives used in the fracturing process, and heavy metals such as barium, iron, boron and more.

Alternatives

Seawater Desalination

In urban areas where climate change has threatened long-term water security and reduced rainfall over catchment areas, using reclaimed water for indirect potable use may be superior to other water supply augmentation methods. One other commonly used option is seawater desalination. Recycling wastewater and desalinating seawater may have many of the same disadvantages, including high costs of water treatment, infrastructure construction, transportation, and waste disposal problems. Although the best option varies from region to region, desalination is often superior economically, as reclaimed water usually requires a dual piping network, often with additional storage tanks, when used for nonpotable use.

Greywater Systems

A less elaborate alternative to reclaimed water is a greywater system. Greywater is wastewater that has been used in sinks, baths, showers, or washing machines, but does not contain sewage and has not been treated at the same levels as recycled water. In a home system, treated or untreated greywater may be used to lush toilets or for irrigation. Some systems now exist which directly use greywater from a sink to lush a toilet.

Rainwater Harvesting

Perhaps the simplest option is a rainwater harvesting system. Although there are concerns about the quality of rainwater in urban areas, due to air pollution and acid rain, many systems exist now to use untreated rainwater for nonpotable uses or treated rainwater for direct potable use. Urban design systems which incorporate rainwater harvesting and reduce runoff are known as Water Sensitive Urban Design (WSUD) in Australia, Low Impact Development (LID) in the United States and Sustainable urban drainage systems (SUDS) in the United Kingdom. There are also concerns about rainwater harvesting systems reducing the amount of run-off entering natural bodies of water.

Health Aspects

Reclaimed water is highly engineered for safety and reliability so that the quality of reclaimed water is more predictable than many existing surface and groundwater sources. Reclaimed water is considered safe when appropriately used. Reclaimed water planned for use in recharging aquifers or augmenting surface water receives adequate and reliable treatment before mixing with naturally occurring water and undergoing natural restoration processes. Some of this water eventually becomes part of drinking water supplies.

A water quality study published in 2009 compared the water quality differences of reclaimed/recycled water, surface water, and groundwater. Results indicate that reclaimed water, surface water, and groundwater are more similar than dissimilar with regard to constituents. The researchers tested for 244 representative constituents typically found in water. When detected, most constituents were in the parts per billion and parts per trillion range. DEET (a bug repellant), and Caffeine were found in all water types and virtually in all samples. Triclosan (in anti-bacterial soap & toothpaste) was found in all water types, but detected in higher levels (parts per trillion) in reclaimed water than in surface or groundwater. Very few hormones/steroids were detected in samples, and when detected were at very low levels. Haloacetic acids (a disinfection by-product) were found in all types of samples, even groundwater. The largest difference between reclaimed water and the other waters appears to be that reclaimed water has been disinfected and thus has disinfection by-products (due to chlorine use).

A 2005 study titled "Irrigation of Parks, Playgrounds, and Schoolyards with Reclaimed Water" found that there had been no incidences of illness or disease from either microbial pathogens or chemicals, and the risks of using reclaimed water for irrigation are not measurably different from irrigation using potable water. Studies by the National Academies of Science, the Monterey Regional Water Pollution Control Agency, and others have found reclaimed water to be safe for agricultural use.

Testing Standards

Reclaimed water is not regulated by the Environmental Protection Agency (EPA), but the EPA has developed water reuse guidelines that were most recently updated in 2012. The EPA Guidelines for Water Reuse represents the international standard for best practices in water reuse. The document was developed under a Cooperative Research and Development Agreement between the U.S. Environmental Protection Agency (EPA), the U.S. Agency for International Development (USAID), and the global consultancy CDM Smith. The Guidelines provide a framework for states to develop regulations that incorporate the best practices and address local requirements.

Ongoing wastewater research sometimes raise concerns about pathogens in the water. Many pathogens cannot be detected by currently used tests.

Recent literature also questions the validity of testing for "indicator organisms" instead of pathogens. Nor do present standards consider interactions of heavy metals and pharmaceuticals which may foster the development of drug resistant pathogens in waters derived from sewage.

To address these concerns about the source water, reclaimed water providers use multi-barrier treatment processes and constant monitoring to ensure that reclaimed water is safe and treated properly for the intended end use.

Potable Use

The main health risk for potable use of reclaimed water is the potential for pharmaceutical and other household chemicals or their derivatives (Environmental persistent pharmaceutical pollutants) to persist in this water. This would be of much less concern if the population were to keep their excrement out of the wastewater e.g. via the use of the Urine-diverting dry toilet or systems that treat blackwater separately from greywater.

Environmental Aspects

There is debate about possible health and environmental effects. To address these concerns, A Risk Assessment Study of potential health risks of recycled water and comparisons to conventional Pharmaceuticals and Personal Care Product (PPCP) exposures was conducted by the WateReuse Research Foundation. For each of four scenarios in which people come into contact with recycled water used for irrigation - children on the playground, golfers, and landscape, and agricultural workers - the findings from the study indicate that it could take anywhere from a few years to millions of years of exposure to nonpotable recycled water to reach the same exposure to PPCPs that we get in a single day through routine activities.

Using reclaimed water for non-potable uses saves potable water for drinking, since less potable water will be used for non-potable uses.

It sometimes contains higher levels of nutrients such as nitrogen, phosphorus and oxygen which may somewhat help fertilize garden and agricultural plants when used for irrigation.

The usage of water reclamation decreases the pollution sent to sensitive environments. It can also enhance wetlands, which benefits the wildlife depending on that eco-system. It also helps to stop the chances of drought as recycling of water reduces the use of fresh water supply from underground sources. For instance, The San Jose/Santa Clara Water Pollution Control Plant instituted a water recycling program to protect the San Francisco Bay area's natural salt water marshes.

Costs and Evaluation

The cost of reclaimed water exceeds that of potable water in many regions of the world, where a fresh water supply is plentiful. However, reclaimed water is usually sold to citizens at a cheaper rate to encourage its use. As fresh water supplies become limited from distribution costs, increased population demands, or climate change reducing sources, the cost ratios will evolve also. The evaluation of reclaimed water needs to consider the entire water supply system, as it may bring important value of flexibility into the overall system

History

Storm and sanitary sewers were necessarily developed along with the growth of cities. By the 1840s the luxury of indoor plumbing, which mixes human waste with water and flushes it away, eliminated the need for cesspools. Odor was considered the big problem in waste disposal and to address it, sewage could be drained to a lagoon, or "settled" and the solids removed, to be disposed of separately. This process is now called "primary treatment" and the settled solids are called "sludge."

At the end of the 19th century, since primary treatment still left odor problems, it was discovered that bad odors could be prevented by introducing oxygen into the decomposing sewage. This was the beginning of the biological aerobic and anaerobic treatments which are fundamental to waste water processes.

By the 1920s, it became necessary to further control the pollution caused by the large quantities of human and industrial liquid wastes which were being piped into rivers and oceans, and modern treatment plants were being built in the US and other industrialized nations by the 1930s.

Designed to make water safe for fishing and recreation, the Clean Water Act of 1972 mandated elimination of the discharge of untreated waste from municipal and industrial sources, and the US federal government provided billions of dollars in grants for building sewage treatment plants around the country. Modern treatment plants, usually using oxidation and/or chlorination in addition to primary and secondary treatment, were required to meet certain standards.

Current treatment improves the quality of separated wastewater solids or sludge. The separated water is given further treatment considered adequate for non potable use by local agencies, and discharged into bodies of water, or reused as reclaimed water. In places like Florida, where it is necessary to avoid nutrient overload of sensitive receiving water, reuse of treated or reclaimed water can be more economically feasible than meeting the higher standards for surface water disposal mandated by the Clean Water Act

Examples

Indirect Potable Reuse (IPR)

- Orange County, California

- Pasadena, California

- Singapore (where it is branded as *NEWater*)

- Payson, Arizona

- The Torreele project in the Veurne coastal region of Belgium, which began operating in 2002

- Virginia Occoquan Reservoir - The Upper Occoquan Sewage Authority plant discharges its highly treated output to supply roughly 20% of the inflow into the Occoquan Reservoir, which provides drinking water used by the Fairfax County Water Authority - one of the three major water providers in the Washington, D.C. metropolitan area.

Non-potable Reuse (NPR)

- Austin, Texas

- Caboolture and Maroochy (South East Queensland, Australia) LGA's currently provide reclaimed water for industrial use (primarily capital works). Users must apply for a key to be able to access the compounds in which the outlets are located.

- Clark County, Nevada

- Clearwater, Florida

- Contra Costa County, California

- Melbourne, Australia

- Mount Buller Ski resort uses recycled water for snow making.

- San Antonio operates the largest recycled water system in the United States.

- Sydney, Australia

- Tucson, Arizona

- San Diego, California (San Diego County)

- St. Petersburg, Florida

Proposed

In some places, reclaimed water has been proposed for either potable or non-potable use:

- South East Queensland, Australia (planned for potable use as of late 2010)

- Newcastle, New South Wales, Australia (proposed for non-potable use).

- Canberra, Australian Capital Territory, Australia (proposed in January 2007 as a backup source of potable water)

- Los Angeles, California - By 2019, the Los Angeles Department of Water and Power will build a plant to replenish their groundwater aquifer with purified water in order to deal with the shortage of rain and snow fall, restricted water imports and local groundwater contamination.

Israel

As of 2010, Israel leads the world in the proportion of water it recycles. Israel treats 80% of its sewage (400 billion liters a year), and 100% of the sewage from the Tel Aviv metropolitan area is treated and reused as irrigation water for agriculture and public works. The remaining sludge is currently pumped into the Mediterranean, however a new bill has passed stating a conversion to treating the sludge to be used as manure. Only 20% of the treated water is lost (due to evaporation, leaks, over-flows and seeping). The recycled water allows farmers to plan ahead and not be limited by water shortages. There are many levels of treatment, and many different ways of treating the water—which leads to a big difference in the quality of the end product. The best quality of reclaimed sewage water comes from adding a gravitational filtering step, after the chemical and biological cleansing. This method uses small ponds in which the water seeps through the sand into the aquifer in about 400 days, then is pumped out as clear purified water. This is nearly the same process used in the space station water recycling system, which turns urine and feces into purified drinking water, oxygen and manure.

To add to the efficiency of the Israeli system - the reclaimed sewage water may be mixed with reclaimed sea water (Plans are in action to increase the desalinization program up to 50% of the countries usage by 2013 - 600 billion liters of drinkable sea water a year), along with aquifer water and fresh sweet lake water - monitored by computer to account for the nationwide needs and input. This action reduced the outdated risk of salt and mineral percentages in the water. Plans to implement this overall usage of reclaimed water for drinking are discouraged by the psychological preconception of the public for the quality of reclaimed water, and the fear of its origin. As of today, all the reclaimed sewage water in Israel is used for agricultural and land improvement purposes.

U.S.

The leaders in use of reclaimed water in the U.S. are Florida and California, with Irvine Ranch Water District as one of the leading developers. They were the first district to approve the use of reclaimed water for in-building piping and use in flushing toilets.

In a January 2012 U.S. National Research Council report, a committee of independent experts found that expanding the reuse of municipal wastewater for irrigation, industrial uses, and drinking water augmentation could significantly increase the United States' total available water resources. The committee noted that a portfolio of treatment options is available to mitigate water quality issues in reclaimed water. The report also includes a risk analysis that suggests the risk of exposure to certain microbial and chemical contaminants from drinking reclaimed water is not any higher than the risk from drinking water from current water treatment systems—and in some cases, may be orders of magnitude lower. The report concludes that adjustments to the federal regulatory framework could enhance public health protection and increase public confidence in water reuse.

Australia

As Australia continues to battle the 7–10-year drought, nationwide, reclaimed effluent is becoming a popular option. Two major capital cities in Australia, Adelaide and Brisbane, have already committed to adding reclaimed effluent to their dwindling dams. The former has also built a desalination plant to help battle any future water shortages. Brisbane has been seen as a leader in this trend, and other cities and towns will review the Western Corridor Recycled Water Project once completed. Goulbourn, Canberra, Newcastle, and Regional Victoria, Australia are already considering building a reclaimed effluent process.

European Union

The second largest waste reclamation program in the world is in Spain, where 12% of the nation's waste is treated.

According to an EU-funded study, "Europe and the Mediterranean countries are lagging behind" California, Japan, and Australia "in the extent to which reuse is being taken up." According to the study "the concept (of reuse) is difficult for the regulators and wider public to understand and accept."

Eco-industrial Park

An eco-industrial park (EIP) is an industrial park in which businesses cooperate with each other and with the local community in an attempt to reduce waste and pollution, efficiently share resources (such as information, materials, water, energy, infrastructure, and natural resources), and help achieve sustainable development, with the intention of increasing economic gains and improving environmental quality. An EIP may also be planned, designed, and built in such a way that it makes it easier for businesses to co-operate, and that results in a more financially sound, environmentally friendly project for the developer.

The Eco-industrial Park Handbook states that "An Eco-Industrial Park is a community of manufacturing and service businesses located together on a common property. Members seek enhanced environmental, economic, and social performance through collaboration in managing environmental and resource issues."

View of the Kalundborg Eco-industrial Park

Based on the concepts of industrial ecology, collaborative strategies not only include by-product synergy ("waste-to-feed" exchanges), but can also take the form of wastewater cascading, shared logistics and shipping & receiving facilities, shared parking, green technology purchasing blocks, multi-partner green building retrofit, district energy systems, and local education & resource centres. This is an application of a systems approach, in which designs and processes/activities are integrated to address multiple objectives.

EIPs can be developed as greenfield land projects, where the eco-industrial intent is present throughout the planning, design and site construction phases, or developed through retrofits and new strategies in existing industrial developments.

Examples

"Industrial symbiosis" is a related but more limited concept in which companies in a region collaborate to utilize each other's by-products and otherwise share resources. In Kalundborg, Denmark a symbiosis network links a 1500MW coal-fired power plant with the community and other companies. Surplus heat from this power plant is used to heat 3500 local homes in addition to a nearby fish farm, whose sludge is then sold as a fertilizer. Steam from the power plant is sold to Novo Nordisk, a pharmaceutical and enzyme manufacturer, in addition to a Statoil plant. This reuse of heat reduces the amount thermal pollution discharged to a nearby fjord. Additionally, a by-product from the power plant's sulfur dioxide scrubber contains gypsum, which is sold to a wallboard manufacturer. Almost all of the manufacturer's gypsum needs are met this way, which reduces the amount of open-pit mining needed. Furthermore, fly ash and clinker from the power plant is utilized for road building and cement production.

The industrial symbiosis at Kalundborg was not created as a top-down initiative, but instead evolved gradually. As environmental regulations became stricter, firms were motivated reduce the cost of compliance, and turn their by-products into economic products.

Example of Industrial Symbiosis. Waste steam from a waste incinerator (right) is piped to an ethanol plant (left) where it is used as in input to their production process.

In Canada, eco-industrial parks exist across the country and have enjoyed some success. The best known example is Burnside Park, in Halifax, Nova Scotia. With support from Dalhousie University's Eco-Efficiency Centre, the more than 1,500 businesses have been improving their environmental performance and developing profitable partnerships. Subsequently, two greenfield industrial developments have been started in Alberta: TaigaNova Eco-Industrial Park is in the heart of the Athabasca oil sands, while Innovista Eco-Industrial Park is a gateway to the Rocky Mountains ~300km west of Edmonton.

Other Usage

EIPs also refer to industrial parks where a "green" approach has been taken towards the infrastructure and development of the site. This can include green infrastructure related to Renewable Energy Systems; stormwater, groundwater and wastewater management; road surfaces; and transportation demand management. Green building practices can also be encouraged or mandated

EIPs are often used as a stimulus for economic diversification in the community or region where they are located. Anchor tenants, such as bio-based product manufacturers or waste-to-energy facilities, etc., can attract complementary businesses as suppliers, scavengers/recyclers, service providers, downstream users and other businesses that could benefit from eco-industrial strategies.

Suggested Usage

It is suggested that EIPs be used as a means of growing the renewable energy sector. In the case of a Solar Photovoltaic (PV) Manufacturing plant, an EIP can increase the manufacturing efficiency to make it more economical, while reducing the environmental impact of producing the solar cells. In essence, this assists the growth of the renewable energy industry and the environmental benefits that come with replacing fossil-fuels.

Industrial Ecology

Industrial ecology (IE) is the study of material and energy flows through industrial systems. The global industrial economy can be modelled as a network of industrial processes that extract resources from the Earth and transform those resources into commodities which can be bought and sold to meet the needs of humanity. Industrial ecology seeks to quantify the material flows and document the industrial processes that make modern society function. Industrial ecologists are often concerned with the impacts that industrial activities have on the environment, with use of the planet's supply of natural resources, and with problems of waste disposal. Industrial ecology is a young but growing multidisciplinary field of research which combines aspects of engineering, economics, sociology, toxicology and the natural sciences.

Industrial ecology has been defined as a "systems-based, multidisciplinary discourse that seeks to understand emergent behaviour of complex integrated human/natural systems". The field approaches issues of sustainability by examining problems from multiple perspectives, usually involving aspects of sociology, the environment, economy and technology. The name comes from the idea that the analogy of natural systems should be used as an aid in understanding how to design sustainable industrial systems.

Overview

Industrial ecology is concerned with the shifting of industrial process from linear (open loop) systems, in which resource and capital investments move through the system to become waste, to a closed loop system where wastes can become inputs for new processes.

Much of the research focuses on the following areas:

- material and energy flow studies ("industrial metabolism")
- dematerialization and decarbonization
- technological change and the environment
- life-cycle planning, design and assessment

- design for the environment ("eco-design")

- extended producer responsibility ("product stewardship")

- eco-industrial parks ("industrial symbiosis")

- product-oriented environmental policy

- eco-efficiency

Industrial ecology seeks to understand the way in which industrial systems (for example a factory, an ecoregion, or national or global economy) interact with the biosphere. Natural ecosystems provide a metaphor for understanding how different parts of industrial systems interact with one another, in an "ecosystem" based on resources and infrastructural capital rather than on natural capital. It seeks to exploit the idea that natural systems do not have waste in them to inspire sustainable design.

Along with more general energy conservation and material conservation goals, and redefining commodity markets and product stewardship relations strictly as a service economy, industrial ecology is one of the four objectives of Natural Capitalism. This strategy discourages forms of amoral purchasing arising from ignorance of what goes on at a distance and implies a political economy that values natural capital highly and relies on more instructional capital to design and maintain each unique industrial ecology.

History

Industrial ecology was popularized in 1989 in a *Scientific American* article by Robert Frosch and Nicholas E. Gallopoulos. Frosch and Gallopoulos' vision was "why would not our industrial system behave like an ecosystem, where the wastes of a species may be resource to another species? Why would not the outputs of an industry be the inputs of another, thus reducing use of raw materials, pollution, and saving on waste treatment?" A notable example resides in a Danish industrial park in the city of Kalundborg. Here several linkages of byproducts and waste heat can be found between numerous entities such as a large power plant, an oil refinery, a pharmaceutical plant, a plasterboard factory, an enzyme manufacturer, a waste company and the city itself.

The scientific field Industrial Ecology has grown quickly in recent years. The Journal of Industrial Ecology (since 1997), the International Society for Industrial Ecology (since 2001), and the journal Progress in Industrial Ecology (since 2004) give Industrial Ecology a strong and dynamic position in the international scientific community. Industrial Ecology principles are also emerging in various policy realms such as the concept of the Circular Economy that is being promoted in China. Although the definition of the Circular Economy has yet to be formalized, generally the focus is on strategies such as creating a circular flow of materials, and cascading energy flows. An example of this would be using waste heat from one process to run another process that requires a lower temperature. The hope is that strategy such as this will create a more efficient economy with fewer pollutants and other unwanted by-products.

Principles

One of the central principles of Industrial Ecology is the view that societal and technological systems are bounded within the biosphere, and do not exist outside of it. Ecology is used as a *metaphor* due to the observation that natural systems reuse materials and have a largely closed loop cycling of nutrients. Industrial Ecology approaches problems with the hypothesis that by using similar principles as *natural systems, industrial systems* can be improved to reduce their impact on the natural environment as well. The table shows the general metaphor.

Biosphere	Technosphere
• Environment	• Market
• Organism	• Company
• Natural Product	• Industrial Product
• Natural Selection	• Competition
• Ecosystem	• Eco-Industrial Park
• Ecological Niche	• Market Niche
• Anabolism / Catabolism	• Manufacturing / Waste Management
• Mutation and Selection	• Design for Environment
• Succession	• Economic Growth
• Adaptation	• Innovation
• Food Web	• Product Life Cycle

The Kalundborg industrial park is located in Denmark. This industrial park is special because companies reuse each other's waste (which then becomes by-products). For example, the Energy E2 Asnæs Power Station produces gypsum as a by-product of the electricity generation process; this gypsum becomes a resource for the BPB Gyproc A/S which produces plasterboards. This is one example of a system inspired by the biosphere-technosphere metaphor: in ecosystems, the waste from one organism is used as inputs to other organisms; in industrial systems, waste from a company is used as a resource by others.

Apart from the direct benefit of incorporating waste into the loop, the use of an eco-industrial park can be a means of making renewable energy generating plants, like Solar PV, more economical and environmentally friendly. In essence, this assists the growth of the renewable energy industry and the environmental benefits that come with re-placing fossil-fuels.

IE examines societal issues and their relationship with both technical systems and the environment. Through this *holistic view* , IE recognizes that solving problems must involve understanding the connections that exist between these systems, various aspects cannot be viewed in isolation. Often changes in one part of the overall system can propagate and cause changes in another part. Thus, you can only understand a problem if you look at its parts in relation to the whole. Based on this framework, IE looks at environmental issues with a *systems thinking* approach.

Take a city for instance. A city can be divided into commercial areas, residential areas, offices, services, infrastructures, and so forth. These are all sub-systems of the 'big city' system. Problems can emerge in one sub-system, but the solution has to be global. Let's say the price of housing is rising dramatically because there is too high a demand for housing. One solution would be to build new houses, but this will lead to more people living in the city, leading to the need for more infrastructure like roads, schools, more supermarkets, etc. This system is a simplified interpretation of reality whose behaviors can be 'predicted'.

In many cases, the systems IE deals with are complex systems. Complexity makes it difficult to understand the behavior of the system and may lead to rebound effects. Due to unforeseen behavioral change of users or consumers, a measure taken to improve environmental performance does not lead to any improvement or may even worsen the situation.

Moreover, *life cycle thinking* is also a very important principle in industrial ecology. It implies that all environmental impacts caused by a product, system, or project during its life cycle are taken into account. In this context life cycle includes

- Raw material extraction

- Material processing

- Manufacture

- Use

- Maintenance

- Disposal

The transport necessary between these stages is also taken into account as well as, if relevant, extra stages such as reuse, remanufacture, and recycle. Adopting a life cycle approach is essential to avoid shifting environmental impacts from one life cycle stage to another. This is commonly referred to as problem shifting. For instance, during the re-design of a product, one can choose to reduce its weight, thereby decreasing use of resources. However, it is possible that the lighter materials used in the new product will be more difficult to dispose of. The environmental impacts of the product gained during the extraction phase are shifted to the disposal phase. Overall environmental improvements are thus null.

A final and important principle of IE is its *integrated approach* or *multidisciplinarity*. IE takes into account three different disciplines: social sciences (including economics), technical sciences and environmental sciences. The challenge is to merge them into a single approach.

Tools

People	Planet	Profit	Modeling
• Stakeholder analysis • Strength Weakness Opportunities Threats Analysis (SWOT Analysis) • Ecolabelling • ISO 14000 • Environmental management system (EMS) • Integrated chain management (ICM) • Technology assessment	• Environmental impact assessment (EIA) • Input-output analysis (IOA) • Life-cycle assessment (LCA) • Material flow analysis (MFA) • Substance flow analysis (SFA) • MET Matrix	• Cost benefit analysis (CBA) • Full cost accounting (FCA) • Life cycle costing (LCC)	• Stock and flow analysis • Agent based modeling

Future Directions

The ecosystem metaphor popularized by Frosch and Gallopoulos has been a valuable creative tool for helping researchers look for novel solutions to difficult problems. Recently, it has been pointed out that this metaphor is based largely on a model of classical ecology, and that advancements in understanding ecology based on complexity science have been made by researchers such as C. S. Holling, James J. Kay, and further advanced in terms of contemporary ecology by others. For industrial ecology, this may mean a shift from a more mechanistic view of systems, to one where sustainability is viewed as an emergent property of a complex system. To explore this further, several researchers are working with agent based modeling techniques .

Exergy analysis is performed in the field of industrial ecology to use energy more efficiently. The term *exergy* was coined by Zoran Rant in 1956, but the concept was developed by J. Willard Gibbs. In recent decades, utilization of exergy has spread outside of physics and engineering to the fields of industrial ecology, ecological economics, systems ecology, and energetics.

History of Industrial Ecology

The birth of industrial ecology is commonly attributed to an article devoted to industrial ecosystems, written by Frosch and Gallopoulos, which appeared in a 1989 special issue of Scientific American, but the field's fundamentals appeared much earlier. Industrial Ecology emerged from several ideas and concepts, some of which date back to the 19th century. Industrial ecology as a concept and as a field of scientific research has developed over time since.

Before The 1960S

The term "Industrial Ecology" has been used alongside "Industrial Symbiosis" at least since the 1940s. Economic geography was perhaps one of the first fields to use these terms. For example, in an article published in 1947, George T. Renner refers to "The General Principle of Industrial Location" as a "Law of Industrial Ecology". Briefly stated this is:

Any industry tends to locate at a point which provides optimum access to its ingredients or component elements. If all these component elements be juxtaposed, the location of the industry is predetermined. If, however, they occur widely separated, the industry is so located as to be most accessible to that element which would be the most expensive or difficult to transport and which, therefore, becomes the locative factor for the industry in question.

In the same article the author defines and describes industrial symbiosis:

Often the location of an industry cannot be fully understood solely in terms of its locative ingredient elements. There are relationships between industries, sometimes simple, but often quite complex, which enter into and complicate the analysis. Chief among these is the phenomenon of industrial symbiosis. By this is meant the consorting together of two or more of dissimilar industries. Industrial Symbiosis, when scrutinized, is seen to be of two kinds, disjunctive and conjunctive.

It appears that the concept of Industrial Symbiosis was not new for the field of economic geography, since the same categorization is used by Walter G. Lezius in his 1937 article "Geography of Glass Manufacture at Toledo, Ohio", also published in the Journal of Economic Geography.

Used in a different context, the term "Industrial Ecology" is also found in a 1958 paper concerned with the relationship between the ecological impact from increasing urbanization and value orientations of related peoples. The case study is in Lebanon:

The central ecological variable in the present research is ecological mobility, or the movement of men in space. It is patent that modern Industrial Ecology requires more such adaptive mobility than does traditional folk-village organization.

1960s

In 1963, we find the term Industrial Ecology (defined as the "complex ecology of the modern industrial world") being used to describe the social nature and complexity of (and within) industrial systems:

...industrial organisations are social rather than mechanical systems. A firm is not only a working organisation with a working purpose. It is rather a community with its own 'politics', in so far as it is involved in problems concerned with the proper distribution of power between individuals and groups of individuals and with questions of individual and group prestige, influence, status and standing... [and he concludes that] the understanding which the student of management is expected to gain is no less than the attainment of insight into an Industrial Ecology of great complexity.

In 1967, the President of the American association for the advancement of science writes in "The experimental city" that "There are examples of industrial symbiosis where one industry feeds off, or at least neutralizes, the wastes of another..." The same author in 1970 talks about "The Next Industrial Revolution" The concept of material and energy sharing and reuse is central to his proposal for a new industrial revolution and he cites agro-industrial symbiosis as a practical way for achieving this:

The object of the next industrial revolution is to ensure that there will be no such thing as waste, on the basis that waste is simply some substance that we do not vet have the wit to use... The next industrial revolution is this generating of a huge new [industry that]... will not produce products, it will rather reprocess the things we call wastes so they may be reproduced in the factories into the things we need... Having the city near the rural area will enable waste heat to be used to speed up the biological processes of treating the organic wastes before they go back into the land. This might end in an elegant arrangement-the power plants located close enough to the center of use, to the people who need the power, but also, within the economics, close enough to the agriculture lands so that the waste heat may be used there. This is an example of agro-industrial symbiosis, if you like to call it that.

In these early articles, "Industrial Ecology" is used in its literal sense - as a system of interacting industrial entities. The relation to natural ecosystems (through either metaphor or analogy) is not explicit. Industrial Symbiosis on the other hand, is already clearly defined as a type of industrial organization, and the term symbiosis is borrowed from the ecological sciences to describe an analogous phenomenon in industrial systems.

1970s

Industrial Ecology has been a research subject of the Japan Industrial Policy Research Institute since 1971. Their definition of Industrial Ecology is "research for the prospect of dynamic harmonization between human activities and nature by a systems approach based upon ecology (JIPRI, 1983)". This programme has resulted to a number of reports that are available only in Japanese.

One of the earliest definitions of Industrial Ecology was proposed by Harry Zvi Evan at a seminar of the Economic Commission of Europe in Warsaw (Poland) in 1973 (an article was subsequently published by Evan in the Journal for International Labour Review in 1974 vol. 110 (3), pp. 219–233). Evan defined Industrial Ecology as a systematic analysis of industrial operations including factors like: Technology, environment, natural resources, bio-medical aspects, institutional and legal matters as well as the socio-economic aspects.

In 1974 the term of Industrial Ecology is perhaps for the first time associated with a cyclical production mode (rather than a linear one, resulting to waste). In this article, the necessity for a transition to an "open-world Industrial Ecology", is used as argument for the need to establish lunar industries:

Low living standards provide one strong motive for most developing countries to increase their productivity and grow economically. Population increase (while it lasts) is a still more powerful driver for increased world consumption. Thus the pressure on resources will continue to grow. Instead of deploring it, we better grow with it. Only through transition to an open-world Industrial Ecology - which includes both benign industrial revolution on Earth and extraterrestrial industrialization - can the present apparent limits to growth be overcome.

Many elements of modern Industrial Ecology were commonplace in the industrial sectors of the former Soviet Union. For example, "kombinirovanaia produksia" (combined production) was present from the earliest years of the Soviet Union and was instrumental in shaping the patterns of Soviet industrialization. "Bezotkhodnoyi tekhnologii" (waste-free technology) was introduced in the final decades of the USSR as a way to increase industrial production while limiting environmental impact. Fiodor Davitaya, a Soviet scientist from the Republic of Georgia, described in 1977 the analogy relating industrial systems to natural systems as a model for a desirable transition to cleaner production:

Nature operates without any waste products. What is rejected by some organisms provides food for others. The organisation of industry on this principle—with the waste products of some branches of industry providing raw material for others—means in effect using natural processes as a model, for in them the resolution of all arising contradictions is the motive force of progress.

1980s

By the 80s Industrial Ecology was already "promoted" to a research subject, which several institutes around the globe embraced. In a 1986 article published in the Journal of Ecological Modeling, there is a full description of Industrial Ecology and the analogy to natural ecosystems is clearly stated:

The structure and inner-working of an industrial society resemble those of a natural ecosystem. The concepts in ecology such as habitat, succession, trophic level, limiting factors and community metabolism can also apply to the study of the ecology of an industrial society. For instance, an industry in a society may grow or decline as a consequence of dynamic changes in exogenous limiting resources and in the hierarchical and/or metabolic structure of that society. When studying the ecology of an industrial society (henceforth termed 'Industrial Ecology'), these concepts and methodologies employed in ecosystems analyses are useful.

In fact, in the above article there is an attempt to model an "industrial ecological system". The model is composed of seven major sections: industry, population, labor force, living state, environment and pollution, general health, and occupational health. Notice the rough similarity with Evan's factors as stated in the above section.

During the 80s the emergence of another related term, "industrial metabolism", is observed. The term is used as a metaphor for the organization and functioning of industrial activity. In an article defending the "biological modulation of terrestrial carbon cycle", the author includes an extraordinary parenthetical note:

Parenthetically, it should be noted that it is an intrinsic property of life to proliferate exponentially until the encounter of limits set by (1) the availability of biologically utilizable reducing power, or (2) the exhaustion of some critical nutrient, or (3) an autotoxic effect imposed by life on its own environment. These limits are universal, applying to microbial ecosystems as well as to the population dynamics of a seemingly unrestricted biological superdominant such as Homo Sapiens (here, the ultimate limit is likely to be placed by an autotoxic effect exerted by the "extrasomatic" (industrial) metabolism of the human race).

1989 – Decisive Articles

In 1989 two articles were released that played a decisive role in the history of industrial ecology. The first one was titled "Industrial Metabolism" by Robert Ayres. Ayres essentially lays the foundations of Industrial Ecology, although the term is not to be found in this article. In the appendix of the article he includes "a theoretical exploration of the biosphere and the industrial economy as material-transformation systems and lessons that might be learned from their comparison". He proposes that:

We may think of both the biosphere and the industrial economy as systems for the transformation of materials. The biosphere as it now exists is nearly a perfect system for recycling materials. This was not the case when life on earth began. The industrial system of today resembles the earliest stage of biological evolution, when the most primitive living organisms obtained their energy from a stock of organic molecules accumulated during prebiotic times. It is increasingly urgent for us to learn from the biosphere and modify our industrial metabolism, the energy - and value - yielding process essential to economic development... we should not only postulate, but indeed endorse, a long-run imperative favoring an industrial metabolism that results in reduced extraction of virgin materials, reduced loss of waste materials, and increased recycling of useful ones.

The term "Industrial Ecology" gains mainstream attention later the same year (1989) through a "Scientific American" article named "Strategies for Manufacturing". In this article, R.Frosch and N.Gallopoulos wonder "why would not our industrial system behave like an ecosystem, where the wastes of a species may be resource to another species? Why would not the outputs of an industry be the inputs of another, thus reducing use of raw materials, pollution, and saving on waste treatment?"

This vision gave birth to the concept of the Eco-industrial Park, the industrial complex that is governed by Industrial Ecology principles. A notable example resides in a Danish industrial park in the city of Kalundborg. There, several linkages of byproducts and waste heat can be found between numerous entities such as a large power plant, an oil refinery, a pharmaceutical plant, a plasterboard factory, an enzyme manufacturer, a waste company and the city itself.

Frosch's and Gallopoulos' thinking was in certain ways simply an extension of earlier ideas, such as the efficiency and waste-reduction thinking annunciated by Buckminster Fuller and his students (e.g., J. Baldwin), and parallel ideas about energy cogeneration, such as those of Amory Lovins and the Rocky Mountain Institute.

1990s

In 1991, C. Kumar Patel organized a seminal colloquium on Industrial Ecology, held on May 20 and 21, 1991, at the National Academy of Sciences in Washington D.C. The papers were later published in the Proceedings of the National Academy of Sciences USA, and they form an excellent reference on Industrial Ecology. Papers include

- "Industrial Ecology: Concepts and Approaches"

- "Industrial Ecology: A Philosophical Introduction"

- "The Ecology of Markets,"

- "Industrial Ecology: Reflections on a Colloquium"

All twenty three papers are available online.

21St Century

The scientific field Industrial Ecology has grown fast in recent years. The Journal of Industrial Ecology (since 1997), the International Society for Industrial Ecology (since 2001), and the journal Progress in Industrial Ecology (since 2004) give Industrial Ecology a strong and dynamic position in the international scientific community. Industrial Ecology principles are also emerging in various policy realms such as the concept of the Circular Economy that is being promoted in China. Although the definition of the Circular Economy has yet to be formalized, generally the focus is on strategies such as creating a circular flow of materials, and cascading energy flows. An example of this would be using waste heat from one process to run another process that requires a lower temperature. This maximizes the efficiency of exergy use. The hope is that strategy such as this will create a more efficient economy with fewer pollutants and other unwanted by products.

References

- Water Reuse: Potential for Expanding the Nation's Water Supply through Reuse of Municipal Wastewater. National Research Council. 2012. ISBN 978-0-309-25749-7.

- Helgeson, Tom (2009). A Reconnaissance-Level Quantitative Comparison of Reclaimed Water, Surface Water, and Groundwater. Alexandria, VA: WateReuse Research Foundation. p. 141. ISBN 978-1-934183-12-0.

- Crook, James (2005). Irrigation of Parks, Playgrounds, and Schoolyards: Extent and Safety. Alexandria, VA: WateReuse Research Foundation. p. 60. ISBN 0-9747586-3-9.

- "Water Recycling and Reuse: The Environmental Benefits/". US Environment Protection Agency. 23 February 2016. Retrieved 22 August 2016.

- Zhang, S.X.; V. Babovic (2012). "A real options approach to the design and architecture of water supply systems using innovative water technologies under uncertainty" (PDF). Journal of Hydroinformatics.

- Levine, Audrey D.; Takashi Asano (1 June 2004). "Peer Reviewed: Recovering Sustainable Water from Wastewater". Environmental Science & Technology. 45: 203A. doi:10.1021/es040504n. Retrieved 20 March 2012.

- Ong, Hian Hai; Ryck, Luc De, "Best Sourcing Approach Keeps Water Production Costs Down", Water & Wastewater International: 13, retrieved 22 March 2012

- "Water Reuse: Potential for Expanding the Nation's Water Supply through Reuse of Municipal Wastewater (2012) : Division on Earth and Life Studies". Retrieved 12 March 2016.

- NEWater FAQ, accessed 8 January 2007; Orange County Water District's Groundwater Replenishment System, accessed 9 September 2011

- "Monterey Wastewater Reclamation Study for Agriculture, Final Report". Monterey Regional Water Pollution Control Agency. Retrieved 18 October 2011.

Permissions

All chapters in this book are published with permission under the Creative Commons Attribution Share Alike License or equivalent. Every chapter published in this book has been scrutinized by our experts. Their significance has been extensively debated. The topics covered herein carry significant information for a comprehensive understanding. They may even be implemented as practical applications or may be referred to as a beginning point for further studies.

We would like to thank the editorial team for lending their expertise to make the book truly unique. They have played a crucial role in the development of this book. Without their invaluable contributions this book wouldn't have been possible. They have made vital efforts to compile up to date information on the varied aspects of this subject to make this book a valuable addition to the collection of many professionals and students.

This book was conceptualized with the vision of imparting up-to-date and integrated information in this field. To ensure the same, a matchless editorial board was set up. Every individual on the board went through rigorous rounds of assessment to prove their worth. After which they invested a large part of their time researching and compiling the most relevant data for our readers.

The editorial board has been involved in producing this book since its inception. They have spent rigorous hours researching and exploring the diverse topics which have resulted in the successful publishing of this book. They have passed on their knowledge of decades through this book. To expedite this challenging task, the publisher supported the team at every step. A small team of assistant editors was also appointed to further simplify the editing procedure and attain best results for the readers.

Apart from the editorial board, the designing team has also invested a significant amount of their time in understanding the subject and creating the most relevant covers. They scrutinized every image to scout for the most suitable representation of the subject and create an appropriate cover for the book.

The publishing team has been an ardent support to the editorial, designing and production team. Their endless efforts to recruit the best for this project, has resulted in the accomplishment of this book. They are a veteran in the field of academics and their pool of knowledge is as vast as their experience in printing. Their expertise and guidance has proved useful at every step. Their uncompromising quality standards have made this book an exceptional effort. Their encouragement from time to time has been an inspiration for everyone.

The publisher and the editorial board hope that this book will prove to be a valuable piece of knowledge for students, practitioners and scholars across the globe.

Index

www.ingramcontent.com/pod-product-compliance
Lightning Source LLC
Chambersburg PA
CBHW061931190326

41458CB00009B/2713